MICROELECTRONIC SYSTEMS 2

A Practical Approach

MICROELECTRONIC SYSTEMS 2

A Practical Approach

W. Ditch B.Sc. Cert. Ed.

Lecturer
Wearside College, Sunderland

Edward Arnold
A member of the Hodder Headline Group
LONDON   SYDNEY   AUCKLAND

For my wife, Julie

© 1995 W. Ditch

First published in Great Britain 1995

British Library Cataloguing in Publication Data

Ditch, W.
    Microelectronic Systems 2 – A Practical Approach
    I. Title
    621.381

    ISBN 0–340–61442–0

Printed and bound in Great Britain for Edward Arnold, a division
of Hodder Headline PLC, 338 Euston Road, London NW1 2BH by
the Bath Press, Avon.

Contents

Preface

This book covers objectives from the BTEC unit *Microelectronic Systems* (U86/333) at level NIII. A previously published companion volume, by the same author, covers the level NII objectives from the same unit.

The text should also be suitable for students studying the *Advanced GNVQ* unit in *Microelectronic Systems* (which was developed for BTEC by the author). Those following other relevant programmes of study including City and Guilds 224, A' level Electronic Systems or first year degrees in Engineering will also find the book to be of interest, as will microelectronics enthusiasts in general.

The text builds on the experience gained with the Z80 and 6502 microprocessors in the previous volume, while introducing more recent developments, such as 16 bit microprocessors and microcontrollers. The first chapter introduces basic concepts related to microprocessor based system design including bus signals, address decoding, control signals and memory system design. The internal operation of commonly available memory devices is also considered with reference to manufacturer's data sheets.

Programming and interfacing experience is further developed by the introduction of interrupt signals and typical microprocessor support ICs. A range of practical applications is also considered including multiplexed LED displays, analogue signal input and output, d.c. and stepper motor control, opto-isolators and a.c. power switching using triacs.

A total of thirty-five practical exercises is included throughout the text, providing reinforcement to the theory as well as ideas for classroom activity or experimentation.

Each chapter has clearly defined aims and those chapters directly related to the BTEC unit have a summary of important points at the end, together with examination-style questions. Answers to these problems are given in a later chapter.

As *integrative assignments* and *common skills* are both compulsory parts of *National* level BTEC courses, two possible integrative assignments are included, together with suitable common skills claim forms. Guidance is also given on the more recently introduced Advanced GNVQ unit in Microelectronic Systems, and on the possible claiming of mandatory GNVQ-related *core skills*.

Acknowledgements

The author would like to thank Texas Instruments who have kindly given permission for the inclusion of extracts from their TTL and MOS Memory databooks.

1   Microprocessor Based Systems

Aims

When you have completed this chapter you should be able to:

1   Understand the function of a microprocessor and the other component parts of a microcomputer.
2   Draw a block diagram of a typical microcomputer including the microprocessor, RAM, ROM, input/output ports and interconnecting buses.
3   Analyse the function and timing of typical control bus signals
4   Recognise the need for address decoding circuitry with associated device enable signals, tri-state outputs and buffers.
5   Interpret schematic diagrams showing the interconnections between the microprocessor, memory and input/output sections.

Introduction

The following section assumes a basic knowledge of microcomputer architecture, which was covered in the preceding volume *Microelectronic Systems 1– A Practical Approach*. Readers who experience difficulty with the concepts presented here are advised to refer to the more detailed coverage given in the first volume!

Microcomputer Architecture

By definition, a *microprocessor* is the *central processing unit* or *CPU* of a *microcomputer*. The microprocessor contains most of the processing elements of a digital computer, with the exception of memory, input/output and miscellaneous logic circuitry.

The microprocessor contains a number of *registers* which may be thought of as latches or special purpose memory locations. These registers may be used to hold actual data (*data registers*), or the address of a data item (*address registers*). The width of address and data registers is normally related to the width of external microprocessor buses.

With an *8 bit microprocessor*, the *data bus* is normally 8 bits wide, while the *address bus* is 16 bits wide. The data bus is used to transfer data to or from the microprocessor via a memory location or an input/output port, and is thus a bi-directional bus. The address bus selects a single memory location or input/output port address for reading or writing. A miscellaneous group of signals called the *control bus* is used to define the direction of data transfer, as well as its timing.

16, 32 and 64 bit microprocessors have more recently become available which, as well as having an increased data bus width, are also capable of directly addressing much larger quantities of memory. For example the 32 bit *Intel 80486* microprocessor has a 32 bit address bus, allowing the microprocessor to directly address 2^{32} memory locations or four *gigabytes* of memory (1 gigabyte = 1024 megabytes).

Figure 1.1 shows a block diagram of a typical 8 bit microcomputer including the microprocessor, ROM, RAM and input/output sections. It can be seen that the address, data and control buses are used to interconnect each section of the microprocessor.

Two types of memory are normally found in a microprocessor based system. These are *random access memory (RAM)* and *read only memory (ROM)*. RAM is normally used to hold temporary data or transient programs and its content is lost when the power supply is removed. It is therefore said to be *volatile*. ROM on the other hand cannot be altered by the user and contains software and data which must be permanently present in the microcomputer's memory. ROM is a *non-volatile* form of storage which is often used to hold the *operating system* or *monitor* programs. In this case the memory range allocated to the ROM is carefully chosen to coincide with the microprocessor's *reset vector*, ensuring that the ROM-based software is automatically executed after the application of power or a *reset* condition.

The *input/output* section is responsible for communication between the microprocessor and external devices, such as keyboards, displays and other peripherals. Its purpose is to simplify the connection of these devices to the computer by the use of latches, shift registers, conversion between analogue and digital signal types, signal conditioning, and special purpose integrated circuits.

Some microprocessors, such as the 6502, treat input/output ports in exactly the same way as memory locations, this being known as *memory mapped input/output*. Other microprocessors including the Z80 use different control signals when dealing with memory and input/output ports. This use of special purpose signals is referred to as *IO mapped input/output*.

IO mapped input/output allows the Z80 microprocessor to address 64K of memory as well as an additional 256 input/output port locations. The design of address decoding circuitry related to input/output ports is also simplified, as only half of the address bus is used to specify the port address.

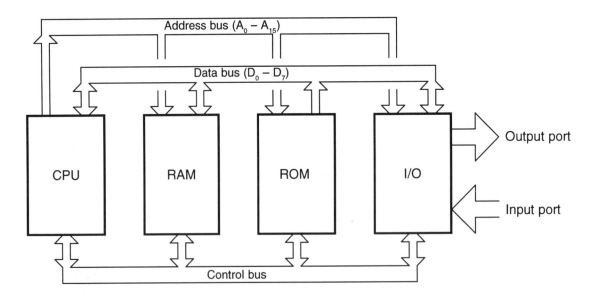

Fig. 1.1 Microcomputer block diagram.

The 6502 Microprocessor

A block diagram of the 6502 microprocessor is given in figure 1.2, showing the registers available to the programmer, as well as the external bus connections.

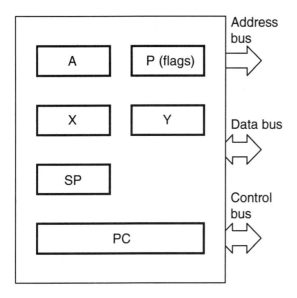

Fig. 1.2 6502 block diagram.

The 6502 has a 16 bit address bus and an 8 bit data bus, which is typical for an 8 bit microprocessor. Table 1.1 lists the names and functions of the internal registers.

Register	Function
Accumulator	An 8 bit general purpose data register, used to hold the result of most arithmetic and logical operations.
X, Y	General purpose data registers, also used in indexed addressing.
P	The processor status register or flags register, which contains information regarding the current state of the microprocessor.
SP	The stack pointer register, used by subroutines and interrupt service routines.
PC	The program counter register which points to the next instruction in the program and is repeatedly incremented during program execution.

Table 1.1 6502 programmer's registers.

As can be seen from figure 1.3, the 6502 is supplied as a 40 pin *dual in line* (*DIL*) package.

Considering the address, data and control buses mentioned earlier, it can be seen that pins A_{15}–A_0 and D_7–D_0 represent the address and data buses respectively. The control bus includes signals such as R/$\overline{\text{W}}$ (Read/$\overline{\text{Write}}$), which controls the direction of data transfer across the data bus, the microprocessor's two phase clock (ϕ_1 and ϕ_2), and the SYNC output which indicates that an op. code fetch is in progress.

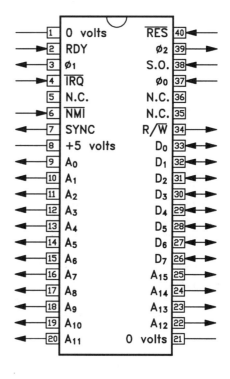

Fig. 1.3 6502 microprocessor pin layout.

The timing of the above mentioned address, data and control signals is extremely important, particularly when interfacing the microprocessor to devices such as memories. Figure 1.4 shows the timing of a 6502 memory read cycle, assuming a 1 MHz oscillator frequency.

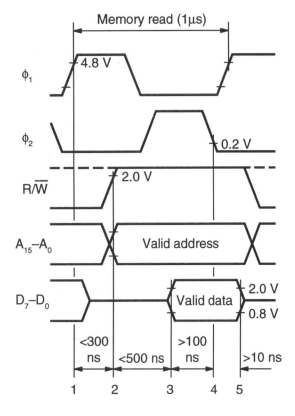

Fig. 1.4   6502 memory read cycle.

The above timing diagram is similar to an op. code fetch cycle, except that the SYNC line is inactive. A memory read cycle may be understood by considering five important points on the diagram.

1   The cycle begins on the rising edge of ϕ_1.
2   Shortly after ϕ_1 all control signals and the address bus are correct. In this case the R/$\overline{\text{W}}$ line goes high, indicating a memory read operation.
3   The external memory device must respond with valid data by this point (500 ns).

4   The information on the data bus is allowed to stabilise and on the falling edge of ϕ_2 the data is latched by the microprocessor.
5   Shortly after being read by the microprocessor, the information on the data bus becomes uncertain.

An important conclusion which can be drawn from figure 1.4, is that the external memory must respond with valid data within 500 ns (0.5 µs) of all address and control signals becoming valid. This is normally referred to as the *access time* and any memory device which cannot satisfy this criteria is said to be incompatible with the microprocessor.

Figure 1.5 shows the timing diagram for a memory write operation.

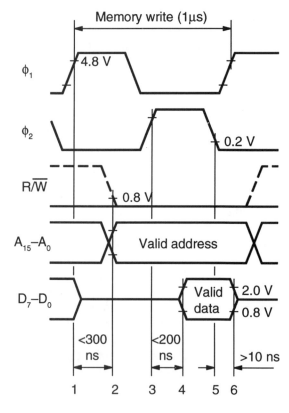

Fig. 1.5   6502 memory write cycle.

Figure 1.5 may be understood by considering six important points on the diagram.

1 The cycle begins on the rising edge of ϕ_1.
2 Shortly after ϕ_1 all control and address signals are valid. In this case the R/$\overline{\text{W}}$ line goes low to indicate a memory write operation.
3 The rising edge of ϕ_2 is used as a timing reference for the availability of valid data (which actually becomes valid at point 4).
4 The microprocessor produces valid data on the data bus, no more than 200 ns after point 3.
5 On the falling edge of ϕ_2 the selected memory device latches the information on the data bus.
6 Shortly after being read, the data bus once again becomes invalid.

The Z80 Microprocessor

A block diagram of the Z80 microprocessor is shown in figure 1.6, showing the registers available to the programmer, as well as the external bus connections.

The Z80 has a 16 bit address bus and an 8 bit data bus, which is typical for an 8 bit microprocessor.

Notice that the Z80 has a main register set and an alternate register set, which may be selected by the use of the appropriate register exchange instructions. This provides a rapid method of *context switching* which, as will be seen later, is particularly suited to the handling of non-maskable interrupts.

Table 1.2 (overleaf) list the names and functions of each register.

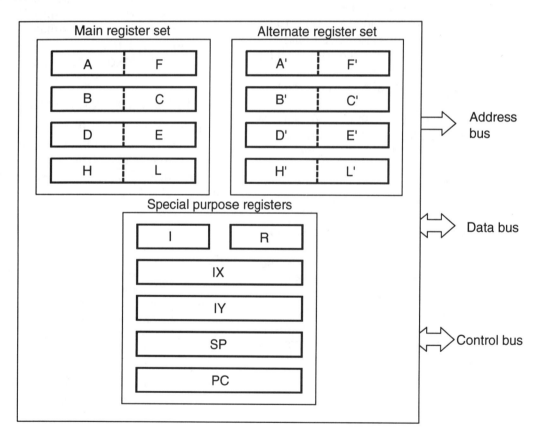

Fig. 1.6 Z80 block diagram.

Register	Function
A	The accumulator. An 8 bit data register used for temporary storage of data, and to hold results produced by the arithmetic and logic unit.
F	The flags register. An 8 bit register containing a series of individual bits or flags which are either set or cleared based on the state of the CPU.
B, C	General purpose registers. These may be used as two separate 8 bit registers or combined to form a 16 bit register pair (the BC register pair).
D, E	General purpose registers. These may be used as two separate 8 bit registers or combined to form a 16 bit register pair (the DE register pair).
H, L	General purpose registers. These may be used as two separate 8 bit registers or combined to form a 16 bit register pair (the HL register pair).
I	The interrupt vector register. An 8 bit register which contains the most significant byte of a 16 bit interrupt vector.
R	The memory refresh register. An 8 bit register used by the Z80's on board dynamic RAM refresh circuitry.
IX	The IX index register. A 16 bit register allowing easy access to lists of data stored in memory .
IY	The IY index register. A 16 bit register providing similar facilities to the IX index register.
SP	The stack pointer. A special purpose register used to control an area of memory known as the stack.
PC	The program counter. A 16 bit register which contains the address of the next instruction in the program. It is repeatedly incremented as the program executes.

Table 1.2   Z80 registers and functions.

The Z80 is supplied as a 40 pin *dual in line* (DIL) package, as can be seen from figure 1.7.

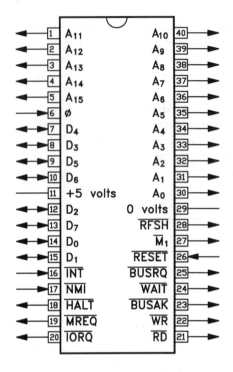

Fig. 1.7   Z80 microprocessor pin layout.

Examining figure 1.7, it can be seen that pins A_{15}–A_0 and D_7–D_0 represent the address and data buses respectively. The control bus includes signals such as $\overline{MREQ}$, $\overline{IORQ}$, $\overline{RD}$, $\overline{WR}$, $\overline{RFSH}$, $\overline{M}_1$ and ϕ. Table 1.3 lists the function of each of these control signals.

Pin	Function
$\overline{MREQ}$	Memory request. Read or write from a memory location.
$\overline{IORQ}$	Input/output request. Read or write from an input/output port.
$\overline{RD}$	Read. Read from memory or I/O port.
$\overline{WR}$	Write. Write to memory or I/O port.
$\overline{RFSH}$	Refresh. Dynamic RAM refresh signal.
$\overline{M}_1$	Machine cycle one (op. code fetch).
ϕ	The Z80 single phase clock waveform.

Table 1.3   Common Z80 control bus signals.

The Z80 is capable of directly addressing 64 kilobytes of memory as well as an additional 256 input/output port locations. The $\overline{\text{MREQ}}$ and $\overline{\text{IORQ}}$ control signals, together with $\overline{\text{RD}}$ and $\overline{\text{WR}}$ are used to differentiate between memory and input/output ports, as shown in table 1.4.

Function	Active (low) pins
Read from memory	$\overline{\text{MREQ}}$, $\overline{\text{RD}}$
Write to memory	$\overline{\text{MREQ}}$, $\overline{\text{WR}}$
Read from I/O port	$\overline{\text{IORQ}}$, $\overline{\text{RD}}$
Write to I/O port	$\overline{\text{IORQ}}$, $\overline{\text{WR}}$

Table 1.4 Z80 input/output control signals.

As the above control signals are active low, a memory read operation, for example, would be characterised by $\overline{\text{MREQ}}$ and $\overline{\text{RD}}$ being low (active), while $\overline{\text{IORQ}}$ and $\overline{\text{WR}}$ are high (inactive). Figure 1.8 shows a timing diagram for a Z80 *memory read cycle*, assuming a 4 MHz oscillator frequency.

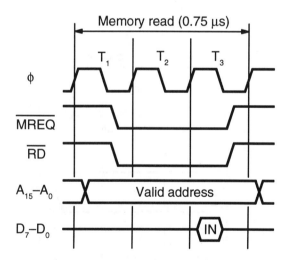

Fig. 1.8 Z80 memory read cycle.

The above memory read cycle is similar in many respects to an op. code fetch, but the $\overline{\text{M}}_1$ signal is inactive and is hence omitted. The memory read cycle is also completed in one cycle fewer than an op. code fetch due to the absence of a dynamic RAM refresh signal.

It can be seen from figure 1.8 that approximately 300 ns elapses between the address and control signals becoming valid, and the data bus being read by the microprocessor. This is normally referred to as the *access time* and any memory device that cannot satisfy this criteria is said to be incompatible with the microprocessor.

Figure 1.9 shows a timing diagram for a Z80 memory write cycle.

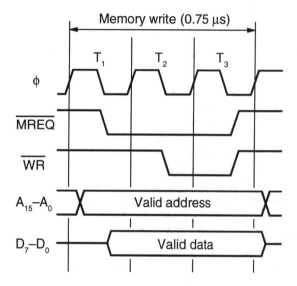

Fig. 1.9 Z80 memory write cycle.

In the case of a memory write cycle, both $\overline{\text{MREQ}}$ and $\overline{\text{WR}}$ are active. The major difference here, compared with a memory read cycle is that the $\overline{\text{WR}}$ line goes low after the falling edge of T_2, rather than T_1. The effect is that the address and data buses are both stable before the $\overline{\text{WR}}$ line becomes active, ensuring that valid data is latched by the memory device.

Input/output read and write operations are similar to memory read or write cycles with the exception that $\overline{\text{IORQ}}$ is used instead of $\overline{\text{MREQ}}$, and that timing requirements are slightly relaxed by the automatic insertion of a wait state in each cycle. In addition, only the lower eight lines of the address bus are significant in the selection the port address. Figure 1.10 (overleaf) gives a timing diagram for successive input/output read and write cycles.

Memory System Design

The 64 kilobyte address space, available to a typical 8 bit microprocessor, is normally divided up into areas which are either RAM, ROM or unused. The quantity of memory required, and its type is determined by the intended application of the computer.

ROM may be used to hold,

- the *operating system* or *monitor*,
- a *boot loader* program,
- permanent data,

while RAM may contain,

- a disc based operating system (loaded by a boot loader),
- transient programs, loaded from disc,
- temporary data.

Having decided on the required memory configuration, a convenient graphical method of displaying this information is to use a *memory map*, as shown by figure 1.11.

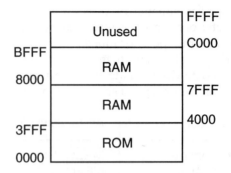

Fig. 1.11   Typical memory map.

A memory map is simply a chart showing the address space of the microprocessor, divided into areas which are either RAM, ROM or unused.

It should also be noted that input/output devices may also appear in this map with the 6502 microprocessor, since all devices are interfaced using *memory mapped input/output* techniques. With the Z80, which uses *IO mapped input/output*, a separate map may be drawn up showing the 256 possible port locations and the 'position' of any input/output devices.

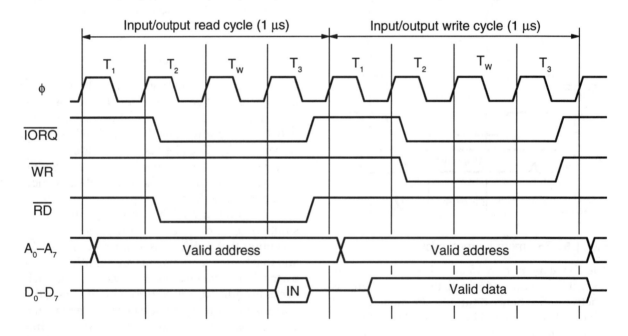

Fig. 1.10   Z80 input/output read and write cycles.

To build a memory circuit satisfying the memory map of figure 1.11, several problems must be overcome.

- Using *tri-state* outputs on all circuits capable of outputting data onto the data bus (which ensures that only one device 'talks' at any time, while other devices 'listen').
- Determining whether *buffering* of the address, data and control buses is required, by examining input and output currents at each device connected to the bus.
- Designing *address decoding* logic to ensure that the appropriate memory device is selected or enabled during a memory read or memory write operation.
- Ensuring that the selected memory devices satisfy any *timing requirements* imposed by the microprocessor, such as the access time.

Tri-State Outputs

The data bus of a typical microcomputer is connected to the microprocessor and a number of memory and input/output devices.

If several devices, capable of outputting data, are connected to a common bus, it is vital that only one device attempts to place information onto the bus at any time. Consider what would happen if two devices attempt to impose different logic levels on a single line, as shown in figure 1.12.

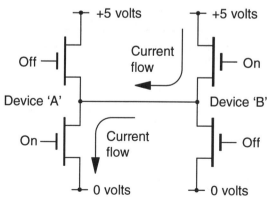

Fig. 1.12   An example of bus contention.

The output circuit is assumed to consist of two transistors connected in series between the supply rails. During normal operation only one of the transistors is switched on, and the conducting transistor provides a low impedance path between the supply rail and the output. For example, if the upper transistor is 'on' the output produces a logic 1.

In figure 1.11, the upper transistor of device 'B' and the lower transistor of device 'A' conduct simultaneously. The result is a large flow of current between the supply rails which will prevent the logic levels present on the bus from being determined, and which may actually destroy the conducting transistors, due to overheating.

The cure to this problem is to place all unused outputs in a high impedance state, as shown in figure 1.13.

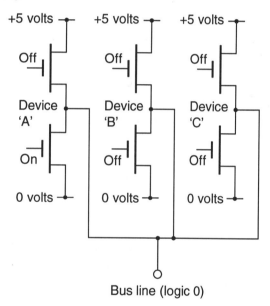

Bus line (logic 0)

Fig. 1.13   A bus using tri-state outputs.

By ensuring that both output transistors are switched-off at the same time, an output may be caused to *float* or enter a *high impedance* state.

In the above example devices A, B and C are connected to a common bus. Device A outputs a logic 0, due its lower transistor conducting. Bus contention is avoided because devices B and C *tri-state* their outputs by disabling both output transistors.

A memory device such as a ROM or RAM, would normally have a special control input whose purpose is to enable or disable the data outputs. When this signal is active it causes the data outputs to be enabled, but when disabled it forces the outputs into a high impedance state (the outputs are said to be *tri-stated*). This signal is not surprisingly called the *output enable* pin, and is shown in figure 1.14.

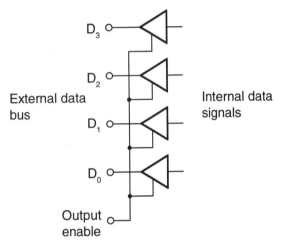

Fig. 1.14 Typical memory output buffer circuit using tri-state devices.

RAM and ROM devices normally have other control inputs, apart from output enable, as shown in figure 1.15.

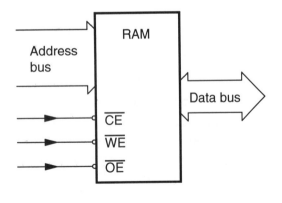

Fig. 1.15(a) Typical RAM block diagram.

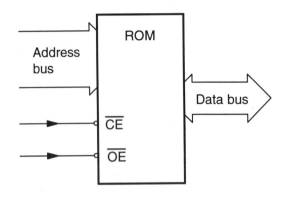

Fig. 1.15(b) Typical ROM block diagram.

The *chip enable* ($\overline{CE}$) input is normally generated directly from the address bus by the address decoding circuitry, as will be seen shortly. When this input is high, the memory device is inactive and the state of the $\overline{WE}$ and $\overline{OE}$ inputs is ignored.

Write enable ($\overline{WE}$) is an input only found on RAM ICs and indicates that the microprocessor is performing a memory write operation. In other words data is transferred from the microprocessor to a memory location specified by the address bus, in conjunction with the address decoding circuitry.

Assuming that the control signals produced by the microprocessor are not directly compatible with those required by the memory IC, it is necessary to design appropriate interface circuitry. With the Z80, the appropriate signals are $\overline{MREQ}$, $\overline{RD}$ and $\overline{WR}$. The output enable ($\overline{OE}$) pin should be pulled low when a memory read occurs, as indicated by $\overline{MREQ}$ and $\overline{RD}$ being simultaneously active. Conversely a memory write operation (to RAM), which is indicated by $\overline{MREQ}$ and $\overline{WR}$ both being active-low, should cause the $\overline{WE}$ pin to pulled low. This is summarised by table 1.5.

Z80 control signals	Memory control input
$\overline{MREQ}$, $\overline{RD}$	$\overline{OE}$
$\overline{MREQ}$, $\overline{WR}$	$\overline{WE}$

Table 1.5 Z80 memory control signals.

Figure 1.16 shows the appropriate Z80 interface logic.

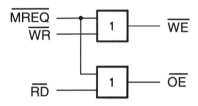

Fig. 1.16 Z80 memory interface logic.

In this case each OR gate produces a *negative logic* AND function. This means that the output function is low only when both of the input variables are at logic 0.

The 6502 microprocessor has a single control line ($R/\overline{W}$) which indicates whether a memory read or memory write operation is in progress. This apparently simple arrangement proves difficult when it comes to designing appropriate memory interface logic. The problem is that using a single control line, there is no way of knowing whether the address and data buses are valid or invalid at a particular instant. The solution to this problem can be seen by examining figures 1.4 and 1.5, which show that whenever ϕ_2 is high, the address bus is valid. It can also be seen that the data bus is always valid on the falling edge of ϕ_2. Figure 1.17 gives the appropriate memory interface logic for the 6502 microprocessor.

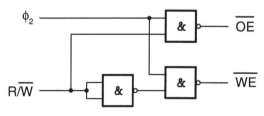

Fig. 1.17 6502 memory interface logic.

The above logic circuit causes the output enable signal to go low when ϕ_2 and $R/\overline{W}$ are both at logic 1 (a memory read operation). A memory write operation, which is characterised by ϕ_2 being at logic 1 while $R/\overline{W}$ is at logic 0, causes the write enable input to go low. (Data is normally latched on the rising edge of $\overline{WE}$).

Buffer Circuits

One problem which frequently occurs in complex bus systems, is that the output device may be incapable of supplying sufficient current to drive all of the inputs connected to the bus. This problem is illustrated by figure 1.18.

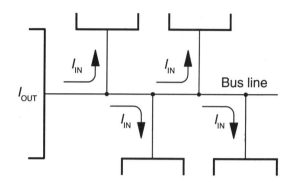

Fig. 1.18 Input and output currents in a typical bus system.

Each of the input currents is supplied by the output device and clearly, as more devices are added to the bus, a point is reached where the output is no longer capable of supplying the required current. This problem may be overcome by inserting a current amplifier or *buffer*, as shown in figure 1.19.

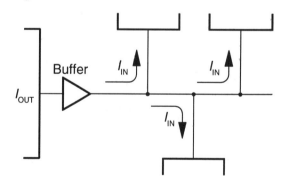

Fig. 1.19 Increasing current drive capacity using a buffer IC.

The above buffer is simply a logic gate with an increased output drive capacity. It allows a larger number of inputs to be driven by the bus, while reducing the loading on the output device.

When deciding whether buffers are required, it is useful to examine the voltage and current characteristics of each input and output connected to the bus.

As will be seen in the next chapter, a wide range of logic families is available, from which microprocessors, memories and other miscellaneous components are constructed. These logic families include TTL (*Transistor Transistor Logic*), CMOS (*Complementary Metal Oxide Semiconductor*) and ECL (*Emitter Coupled Logic*). There are also many 'improved' varieties available, which have particular areas of application such as low power consumption or high operating speed. Table 1.6 lists some commonly available logic families and their major characteristics. (This is not an exhaustive list, since it omits some very high speed logic families).

The parameters given in table 1.6 may be best understood by considering the allowed logic levels on an output, which is connected to a corresponding input, as shown in figure 1.20.

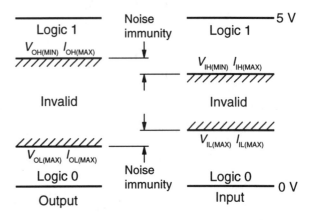

Fig. 1.20   Allowed input/output logic levels.

Parameter		TTL	LSTTL	ALSTTL	CMOS	74HC	74HCT
V_{CC}	(min)	4.75 V	4.75 V	4.75 V	3 V	2 V	4.5 V
	(max)	5.25 V	5.25 V	5.25 V	18 V	6 V	5.5 V
$V_{IL(MIN)}$	(5 V)	0.8 V	0.8 V	0.8 V	1.5 V	1.1 V	0.8 V
$I_{IL(MAX)}$		−1.6 mA	−0.4 mA	−0.2 mA	−0.005 μA	0	0
$V_{IH(MIN)}$	(5 V)	2.0 V	2.0 V	2.0 V	3.5 V	3.15 V	2.0 V
$I_{IH(MAX)}$		40 μA	20 μA	20 μA	0.005 μA	0	0
$V_{OL(MAX)}$	(5 V)	0.4 V	0.4 V	0.4 V	0.4 V	0.4 V	0.4 V
$I_{OL(MAX)}$		16 mA	8 mA	4 mA	0.4 mA	4 mA	4 mA
$V_{OH(MAX)}$	(5 V)	2.4 V	2.7 V	2.7 V	4.5 V	3.7 V	3.7 V
$I_{OH(MAX)}$		−400 μA	−400 μA	−400 μA	−0.4 mA	−4 mA	−0.4 mA
Dissipation per gate		10 mW	2 mW	1 mW	2.5 nW	2.5 nW	2.5 nW
Propagation delay		10 ns	9.5 ns	4 ns	40 ns	9 ns	9 ns
		(Basic 74 series TTL)	(Low power schottky)	(Advanced LS)	(Basic 4000 series CMOS)	(High speed 74 series CMOS)	(HCMOS TTL pin compatible)

Table 1.6   Common logic families and their principal characteristics.

Each logic level is actually specified as a range of voltages, with an invalid zone between the two logic levels. The maximum output current at each logic level is given, together with the worst possible output voltage which may occur. For example a TTL logic 1 output is guaranteed to be greater than 2.4 volts. This means that V_{OH} will be greater than 2.4 V, provided that I_{OH} is less than 400 µA. (Note that the minus sign in the table indicates that current flows out of the gate). The corresponding minimum logic 1 input voltage is specified as being greater than 2.0 volts ($V_{IH} > 2.0$ volts), while the maximum current taken by the input is specified as 40 µA ($I_{IH} < 40$ µA).

There are two important conclusions which may be drawn from the above figures. Firstly a single TTL output is capable of directly driving ten TTL inputs ($I_{OH(MAX)} = -400$ µA while $I_{IH(MIN)} = 40$ µA) in the logic 1 state. A check of the corresponding logic 0 input and output currents also shows that a single output can drive up to ten inputs ($I_{OL(MAX)} = 16$ mA, while $I_{IL(MAX)} = 1.6$ mA). The gate is said to have a *fan out* of ten, meaning that up to ten inputs can be driven from a single output, at either logic level. Secondly, the minimum logic 1 output voltage is 2.4 volts, while any input voltage greater than 2.0 volts will be accepted as a logic 1. This means that up to 0.4 volts of electrical noise could be induced between the input and output without causing a malfunction. The *noise immunity* at logic 1 is said to be 400 mV. From table 1.6 it can be seen that the value of noise immunity at logic 0 is also 0.4 volts ($V_{OL(MAX)} = 0.4$ V, while $V_{IL(MAX)} = 0.8$ V).

As will be seen in chapter 6, there are many sources of electrical noise and interference in a typical microcomputer and its immediate environment. Careful system design is required to minimise these effects. In fact the noise immunity of standard TTL gates is rather poor compared with CMOS, for example. For this reason buffer circuits are often constructed with *hysteresis* at the input, effectively increasing the noise immunity of the gate. This effect is provided by a *Schmitt trigger* input circuit, whose circuit symbol is shown in figure 1.21.

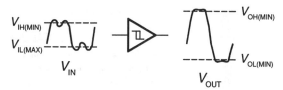

Fig. 1.21   Buffer with Schmitt trigger input.

In the above illustration, a 'noisy' input signal is applied to the input of the buffer and due to the hysteresis of the input circuit, the logic levels at the output are corrected.

Other useful properties of dedicated buffer circuits are that the outputs are capable of supplying much larger currents than standard ICs in their logic family. For example 74 and 74LS buffers have a fan out which is three times as large as normal 74 or 74LS devices, while the fan out of HC and HCT ICs with bus driver outputs is increased by 50%. This means that a TTL buffer is capable of driving 30 inputs of the same logic family, rather than the normal limit of 10 logic gates.

Figure 1.22 gives details of the '240, '241 and '244 buffer ICs, which are available as part of the TTL logic family (reproduced by courtesy of Texas Instruments). Notice in particular the use of Schmitt trigger inputs, tri-state outputs and the increased output current capacity provided by each gate.

**TYPES SN74LS240, SN74LS241, SN74LS244, SN74S240, SN74S241, SN74S244
SN54LS240, SN54LS241, SN54LS244, SN54S240, SN54S241, SN54S244
OCTAL BUFFERS AND LINE DRIVERS WITH 3-STATE OUTPUTS**

REVISED APRIL 1985

- **3-State Outputs Drive Bus Lines or Buffer Memory Address Registers**

- **PNP Inputs Reduce D-C Loading**

- **Hysteresis at Inputs Improves Noise Margins**

description

These octal buffers and line drivers are designed specifically to improve both the performance and density of three-state memory address drivers, clock drivers, and bus-oriented receivers and transmitters. The designer has a choice of selected combinations of inverting and noninverting outputs, symmetrical $\overline{G}$ (active-low output control) inputs, and complementary G and $\overline{G}$ inputs. These devices feature high fan-out, improved fan-in, and 400-mV noise-margin. The SN74LS' and SN74S' can be used to drive terminated lines down to 133 ohms.

The SN54' family is characterized for operation over the full military temperature range of $-55\,^\circ$C to $125\,^\circ$C. The SN74' family is characterized for operation from $0\,^\circ$C to $70\,^\circ$C.

**SN54LS', SN54S' ... J PACKAGE
SN74LS', SN74S' ... DW OR N PACKAGE
(TOP VIEW)**

$1\overline{G}$	1	20 V_{CC}
1A1	2	19 $2\overline{G}/2G^*$
2Y4	3	18 1Y1
1A2	4	17 2A4
2Y3	5	16 1Y2
1A3	6	15 2A3
2Y2	7	14 1Y3
1A4	8	13 2A2
2Y1	9	12 1Y4
GND	10	11 2A1

**SN54LS', SN54S' ... FK PACKAGE
(TOP VIEW)**

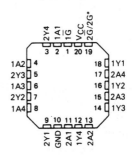

*2G for 'LS241 and 'S241 or $2\overline{G}$ for all other drivers.

schematics of inputs and outputs

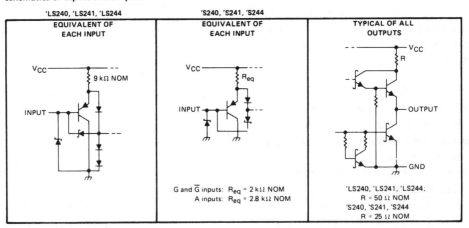

TEXAS INSTRUMENTS

Fig. 1.22 (a)   Octal buffer data sheets (courtesy of Texas Instruments).

TYPES SN74LS240, SN74LS241, SN74LS244, SN74S240, SN74S241, SN74S244
SN54LS240, SN54LS241, SN54LS244, SN54S240, SN54S241, SN54S244
OCTAL BUFFERS AND LINE DRIVERS WITH 3-STATE OUTPUTS

logic symbols

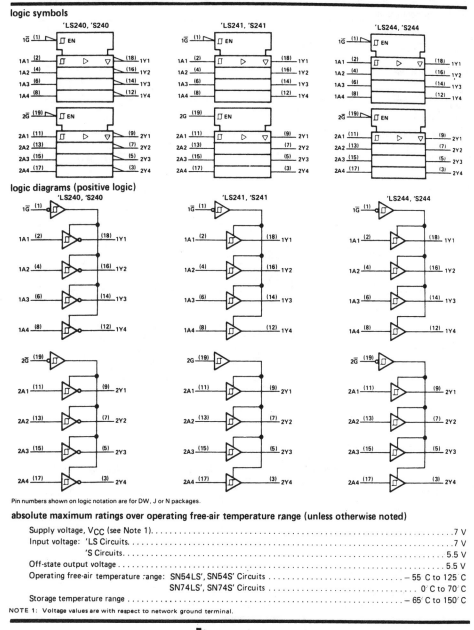

logic diagrams (positive logic)

Pin numbers shown on logic notation are for DW, J or N packages.

absolute maximum ratings over operating free-air temperature range (unless otherwise noted)

Supply voltage, V$_{CC}$ (see Note 1)..7 V

Input voltage: 'LS Circuits...7 V

'S Circuits..5.5 V

Off-state output voltage..5.5 V

Operating free-air temperature range: SN54LS', SN54S' Circuits............................ −55°C to 125°C

SN74LS', SN74S' Circuits................................. 0°C to 70°C

Storage temperature range..−65°C to 150°C

NOTE 1: Voltage values are with respect to network ground terminal.

3-514

TEXAS INSTRUMENTS

Fig. 1.22 (b) Octal buffer data sheets (courtesy of Texas Instruments).

TYPES SN74LS240, SN74LS241, SN74LS244, SN54LS240, SN54LS241, SN54LS244
OCTAL BUFFERS AND LINE DRIVERS WITH 3-STATE OUTPUTS

recommended operating conditions

PARAMETER		SN54LS' MIN	NOM	MAX	SN74LS' MIN	NOM	MAX	UNIT
V_{CC}	Supply voltage (see Note 1)	4.5	5	5.5	4.75	5	5.25	V
V_{IH}	High-level input voltage	2			2			V
V_{IL}	Low-level input voltage			0.7			0.8	V
I_{OH}	High-level output current			−12			−15	mA
I_{OL}	Low-level output current			12			24	mA
T_A	Operating free-air temperature	−55		125	0		70	°C

NOTE 1: Voltage values are with respect to network ground terminal.

electrical characteristics over recommended operating free-air temperature range (unless otherwise noted)

PARAMETER		TEST CONDITIONS[†]		SN54LS' MIN	TYP[‡]	MAX	SN74LS' MIN	TYP[‡]	MAX	UNIT
V_{IK}		V_{CC} = MIN,	I_I = −18 mA			−1.5			−1.5	V
Hysteresis $(V_{T+} - V_{T-})$		V_{CC} = MIN		0.2	0.4		0.2	0.4		V
V_{OH}		V_{CC} = MIN, V_{IH} = 2 V, I_{OH} = −3 mA	V_{IL} = MAX,	2.4	3.4		2.4	3.4		V
		V_{CC} = MIN, V_{IH} = 2 V, I_{OH} = MAX	V_{IL} = 0.5 V,	2			2			
V_{OL}		V_{CC} = MIN, V_{IH} = 2 V, V_{IL} = MAX	I_{OL} = 12 mA			0.4			0.4	V
			I_{OL} = 24 mA						0.5	
I_{OZH}		V_{CC} = MAX, V_{IH} = 2 V,	V_O = 2.7 V			20			20	μA
I_{OZL}		V_{IL} = MAX	V_O = 0.4 V			−20			−20	
I_I		V_{CC} = MAX,	V_I = 7 V			0.1			0.1	mA
I_{IH}		V_{CC} = MAX,	V_I = 2.7 V			20			20	μA
I_{IL}		V_{CC} = MAX,	V_{IL} = 0.4 V			−0.2			−0.2	mA
I_{OS}[§]		V_{CC} = MAX		−40		−225	−40		−225	mA
I_{CC}	Outputs high	V_{CC} = MAX, Output open	All	17	27		17	27		mA
	Outputs low		'LS240	26	44		26	44		
			'LS241, 'LS244	27	46		27	46		
	All outputs disabled		'LS240	29	50		29	50		
			'LS241, 'LS244	32	54		32	54		

† For conditions shown as MIN or MAX, use the appropriate value specified under recommended operating conditions.
‡ All typical values are at V_{CC} = 5 V, T_A = 25°C.
§ Not more than one output should be shorted at a time, and duration of the short-circuit should not exceed one second.

switching characteristics, V_{CC} = 5 V, T_A = 25°C

PARAMETER	TEST CONDITIONS		'LS240 MIN	TYP	MAX	'LS241, 'LS244 MIN	TYP	MAX	UNIT
t_{PLH}	R_L = 667 Ω, See Note 2	C_L = 45 pF,		9	14		12	18	ns
t_{PHL}				12	18		12	18	ns
t_{PZL}				20	30		20	30	ns
t_{PZH}				15	23		15	23	ns
t_{PLZ}	R_L = 667 Ω, See Note 2	C_L = 5 pF,		10	20		10	20	ns
t_{PHZ}				15	25		15	25	ns

NOTE 2: See General Information Section for load circuits and voltage waveforms.

TEXAS INSTRUMENTS

Fig. 1.22 (c) Octal buffer data sheets (courtesy of Texas Instruments).

The buffer ICs of figure 1.22 would be suitable for increasing the current driving capacity of microprocessor signals which are outputs from the microprocessor. These would include the address bus and most of the control bus. In a typical 8 bit microcomputer, three such ICs would be sufficient to buffer all microprocessor output signals.

Figure 1.23 shows typical pin connections to a 74LS244 octal buffer IC, when used to buffer up to eight microprocessor output signals. Notice that pins 1 and 19 are connected together and used as an output enable signal. By connecting this signal to logic 0, the buffered outputs are enabled, while a logic 1 causes the outputs to be tri-stated.

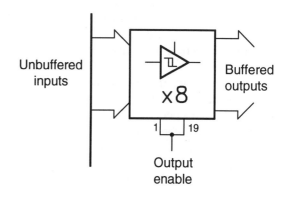

Fig. 1.23 Typical octal buffer pin connections.

Practical Exercise   Schmitt Trigger Circuit

Method

Test Circuit

Using the potentiometer, vary V_{IN} while observing the voltage at the output of the operational amplifier. Notice in particular the input voltages which cause V_{OUT} to change when V_{IN} is,

    a) increased (upper threshold voltage),
    b) decreased (lower threshold voltage).

Comment on the usefulness of this circuit for the recovery of digital signals in electrically 'noisy' environments.

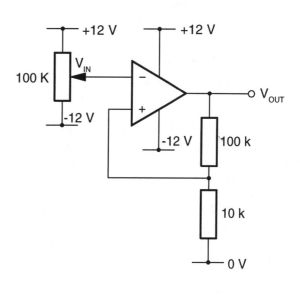

Equipment Required

+12/0/–12 volt d.c. power supply
Prototyping board
Operational amplifier (741 or similar)
100 kΩ resistor
10 kΩ resistor
100 kΩ (linear) potentiometer
2 digital voltmeters

Although the buffer circuit of figure 1.23 is suitable for unidirectional signals, such as those found on the address bus, a different circuit must be used with bi-directional signals such as the data bus.

During a *memory write* operation, the data bus acts as an output from the microprocessor and buffering may be needed on complex systems. The data bus also acts as an input to the microprocessor during *memory read* operations, preventing the use of the previously considered tri-state buffer. The circuit of figure 1.24 overcomes this problem by connecting two buffers 'back to back'.

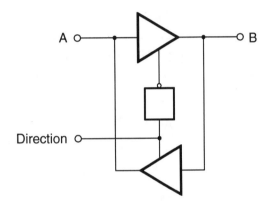

Fig. 1.24   Bi-directional buffer circuit.

In this case, if the *direction* signal is at logic 1, the lower buffer is enabled and data is transferred from right to left (B to A). Conversely, a logic 0 on the direction input enables the upper buffer (via the inverter), causing data to be transferred from left to right. The inverter ensures that only one buffer can be enabled at any time, preventing any contention.

A slight disadvantage of the above circuit is that one of the outputs is always enabled. In microcomputer applications where external devices may take control of the buses, it is useful to be able to completely disable the buffer outputs. This may be achieved by adding an *output enable* circuit, in addition to the direction control, as shown in figure 1.25.

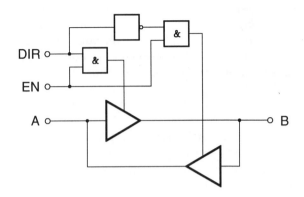

Fig. 1.25   Bi-directional buffer with output enable.

When the *enable* input is at logic 0, both buffers are tri-stated, causing A and B to be electrically isolated. Once enable goes to logic 1, the direction of data transfer may be selected using the *direction* input. The operation of this transmitter/receiver circuit (commonly called a *transceiver*), is summarised in table 1.7.

Enable	Direction	Function
0	0	high impedance
0	1	high impedance
1	0	B to A
1	1	A to B

Table 1.7   Bus transceiver truth table.

Figure 1.26 shows data sheets of the 74LS245 octal (8 input) bus transceiver, which is available as part of the TTL logic family (reproduced by courtesy of Texas Instruments). Once again notice the use of Schmitt trigger inputs to improve noise immunity and the increased current capacity of each output.

TYPES SN74LS245, SN54LS245
OCTAL BUS TRANSCEIVERS WITH 3-STATE OUTPUTS
OCTOBER 1976 - REVISED APRIL 1985

- Bi-directional Bus Transceiver in a High-Density 20-Pin Package

- 3-State Outputs Drive Bus Lines Directly

- PNP Inputs Reduce D-C Loading on Bus Lines

- Hysteresis at Bus Inputs Improve Noise Margins

- Typical Propagation Delay Times, Port-to-Port . . . 8 ns

TYPE	I_{OL} (SINK CURRENT)	I_{OH} (SOURCE CURRENT)
SN54LS245	12 mA	-12 mA
SN74LS245	24 mA	-15 mA

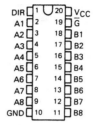

SN54LS245 ... J PACKAGE
SN74LS245 ... DW OR N PACKAGE
(TOP VIEW)

DIR	1	20	V_{CC}
A1	2	19	G
A2	3	18	B1
A3	4	17	B2
A4	5	16	B3
A5	6	15	B4
A6	7	14	B5
A7	8	13	B6
A8	9	12	B7
GND	10	11	B8

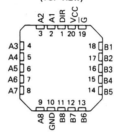

SN54LS245 ... FK PACKAGE
(TOP VIEW)

description

These octal bus transceivers are designed for asynchronous two-way communication between data buses. The control function implementation minimizes external timing requirements.

The devices allow data transmission from the A bus to the B bus or from the B bus to the A bus depending upon the logic level at the direction control (DIR) input. The enable input (G) can be used to disable the device so that the buses are effectively isolated.

The SN54LS245 is characterized for operation over the full military temperature range of −55°C to 125°C. The SN74LS245 is characterized for operation from 0°C to 70°C.

schematics of inputs and outputs

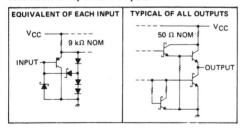

FUNCTION TABLE

ENABLE $\bar{G}$	DIRECTION CONTROL DIR	OPERATION
L	L	B data to A bus
L	H	A data to B bus
H	X	Isolation

H = high level, L = low level, X = irrelevant

TEXAS INSTRUMENTS

Fig. 1.26 (a) Octal bus transceiver data sheets (courtesy of Texas Instruments).

TYPES SN74LS245, SN54LS245
OCTAL BUS TRANSCEIVERS WITH 3-STATE OUTPUTS

logic symbol

logic diagram (positive logic)

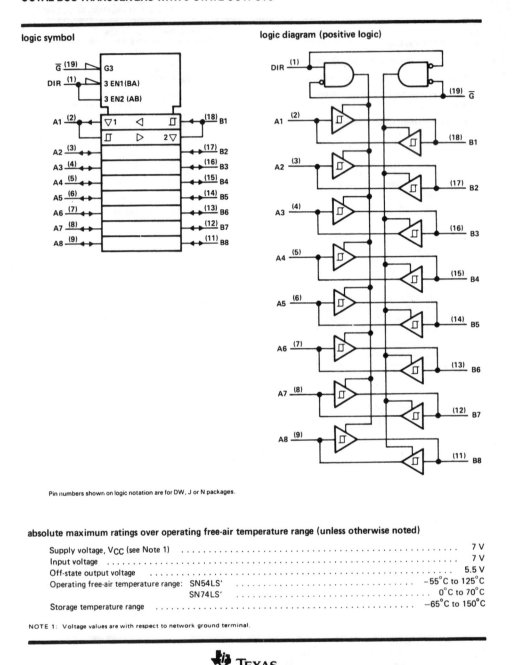

Pin numbers shown on logic notation are for DW, J or N packages.

absolute maximum ratings over operating free-air temperature range (unless otherwise noted)

Supply voltage, V_{CC} (see Note 1)	7 V
Input voltage	7 V
Off-state output voltage	5.5 V
Operating free-air temperature range: SN54LS'	$-55°C$ to $125°C$
SN74LS'	$0°C$ to $70°C$
Storage temperature range	$-65°C$ to $150°C$

NOTE 1: Voltage values are with respect to network ground terminal.

TEXAS INSTRUMENTS

Fig. 1.26 (b) Octal bus transceiver data sheets (courtesy of Texas Instruments).

TYPES SN74LS245, SN54LS245
OCTAL BUS TRANSCEIVERS WITH 3-STATE OUTPUTS

recommended operating conditions

PARAMETER	SN54LS245			SN74LS245			UNIT
	MIN	NOM	MAX	MIN	NOM	MAX	
Supply voltage, V_{CC}	4.5	5	5.5	4.75	5	5.25	V
High-level output current, I_{OH}			−12			−15	mA
Low-level output current, I_{OL}			12			24	mA
Operating free-air temperature, T_A	−55		125	0		70	°C

electrical characteristics over recommended operating free-air temperature range (unless otherwise noted)

PARAMETER		TEST CONDITIONS†		SN54LS245			SN74LS245			UNIT
				MIN	TYP‡	MAX	MIN	TYP‡	MAX	
V_{IH}	High-level input voltage			2			2			V
V_{IL}	Low-level input voltage					0.7			0.8	V
V_{IK}	Input clamp voltage	V_{CC} = MIN,	I_I = −18 mA			−1.5			−1.5	V
	Hysteresis $(V_{T+} - V_{T-})$ A or B input	V_{CC} = MIN		0.2	0.4		0.2	0.4		V
V_{OH}	High-level output voltage	V_{CC} = MIN, V_{IH} = 2 V, $V_{IL} = V_{IL}$ max	I_{OH} = −3 mA	2.4	3.4		2.4	3.4		V
			I_{OH} = MAX	2			2			
V_{OL}	Low-level output voltage	V_{CC} = MIN, V_{IH} = 2 V, $V_{IL} = V_{IL}$ max	I_{OL} = 12 mA			0.4			0.4	V
			I_{OL} = 24 mA						0.5	
I_{OZH}	Off-state output current, high-level voltage applied	V_{CC} = MAX, $\overline{G}$ at 2 V	V_O = 2.7 V			20			20	μA
I_{OZL}	Off-state output current, low-level voltage applied		V_O = 0.4 V			−200			−200	
I_I	Input current at A or B	V_{CC} = MAX,	V_I = 5.5 V			0.1			0.1	mA
	maximum input voltage DIR or $\overline{G}$		V_I = 7 V			0.1			0.1	
I_{IH}	High-level input current	V_{CC} = MAX,	V_{IH} = 2.7 V			20			20	μA
I_{IL}	Low-level input current	V_{CC} = MAX,	V_{IL} = 0.4 V			−0.2			−0.2	mA
I_{OS}	Short-circuit output current¶	V_{CC} = MAX		−40		−225	−40		−225	mA
I_{CC}	Supply current — Total, outputs high	V_{CC} = MAX,	Outputs open		48	70		48	70	mA
	Total, outputs low				62	90		62	90	
	Outputs at Hi-Z				64	95		64	95	

†For conditions shown as MIN or MAX, use the appropriate value specified under recommended operating conditions.
‡All typical values are at V_{CC} = 5 V, T_A = 25°C.
¶Not more than one output should be shorted at a time, and duration of the short-circuit should not exceed one second.

switching characteristics, V_{CC} = 5 V, T_A = 25°C

PARAMETER		TEST CONDITIONS	MIN	TYP	MAX	UNIT
t_{PLH}	Propagation delay time, low-to-high-level output	C_L = 45 pF, R_L = 667 Ω, See Note 2		8	12	ns
t_{PHL}	Propagation delay time, high-to-low-level output			8	12	ns
t_{PZL}	Output enable time to low level			27	40	ns
t_{PZH}	Output enable time to high level			25	40	ns
t_{PLZ}	Output disable time from low level	C_L = 5 pF, R_L = 667 Ω, See Note 2		15	25	ns
t_{PHZ}	Output disable time from high level			15	28	ns

NOTE 2: See General Information Section for load circuits and voltage waveforms.

TEXAS
INSTRUMENTS

3-525

Fig. 1.26 (c) Octal bus transceiver data sheets (courtesy of Texas Instruments).

Figure 1.27 shows the method used to buffer the address and data bus of a typical 8 bit microcomputer (neglecting the buffering of control signals).

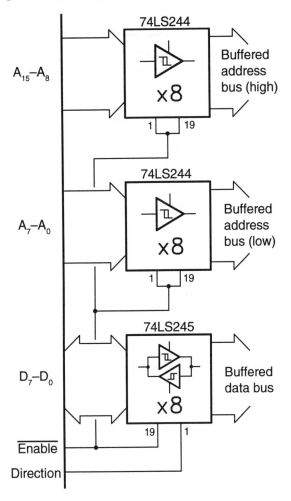

Fig. 1.27 Address and data bus buffering.

The direction of data transfer would be selected by connecting an appropriate microprocessor control signal to the *direction* input (for example $\overline{RD}$ on the Z80 or $R/\overline{W}$ on the 6502).

The microprocessor buses may be isolated from the rest of the microcomputer by outputting a logic 1 on the $\overline{Enable}$ line. This might be required on the Z80 during a *bus acknowledge* cycle, following a *direct memory access* (*DMA*) request. On systems not supporting DMA requests, $\overline{Enable}$ would be directly connected to logic 0.

Address Decoding

As has already been seen, contention between devices connected to the data bus may be avoided by the use of tri-state outputs and suitable enabling logic. *Address decoding* logic is an important part of this process.

The *address space* of a typical 8 bit microprocessor is 64 kilobytes (while more modern 16, 32 and 64 bit microprocessors are capable of addressing several megabytes of memory at least).

Memory devices (whose internal operation is covered in the next chapter), are normally available with memory capacities which are multiples of 1 kilobyte. Typically sizes of memory used in 8 bit systems is therefore 1, 2, 4, 8, 16 or 32 kilobytes, and combinations of these devices must be used to form the required microcomputer memory map.

Once the quantity of memory required by a microcomputer, and its type have been chosen, the address decoding logic may now be designed. The first thing to consider is the *significance* of each address line, as shown in figure 1.28.

		16 kilobytes (C000 - FFFF$_{16}$)
32 kilobytes (8000 - FFFF$_{16}$)		$A_{15} = 1, A_{14} = 1$
		16 kilobytes (8000 - BFFF$_{16}$)
$A_{15} = 1$		$A_{15} = 1, A_{14} = 0$
		16 kilobytes (4000 - 7FFF$_{16}$)
32 kilobytes (0000 - 7FFF$_{16}$)		$A_{15} = 0, A_{14} = 1$
		16 kilobytes (0000 - 3FFF$_{16}$)
$A_{15} = 0$		$A_{15} = 0, A_{14} = 0$

Fig. 1.28 Address bus significance.

The most significant address line divides the address space into two 32 kilobyte blocks, while the four possible combinations of A_{15} and A_{14} each select a 16 kilobyte memory range.

Assuming that 16 kilobyte memory devices are used, address lines A_{15} and A_{14} may be connected directly to the address decoding circuitry. The remaining address lines (A_{13}–A_0) are connected to the address inputs of each memory device.

A *2 to 4 line decoder* IC may be used to provide the required chip enable signals. The decoder has two inputs and four outputs. Based on the four possible combinations of input (00, 01, 10 or 11) the circuit activates one of the four outputs, as shown in table 1.8.

A	B	Y_0	Y_1	Y_2	Y_3
0	0	0	1	1	1
0	1	1	0	1	1
1	0	1	1	0	1
1	1	1	1	1	0

Table 1.8 2 to 4 line decoder truth table.

Figure 1.29 shows the address decoding circuitry for a microcomputer with 16 kilobytes of ROM and 32 kilobytes of RAM. The address range $C000_{16}$ to $FFFF_{16}$ is unused, but the provision of a suitable address select signal simplifies any future memory expansion.

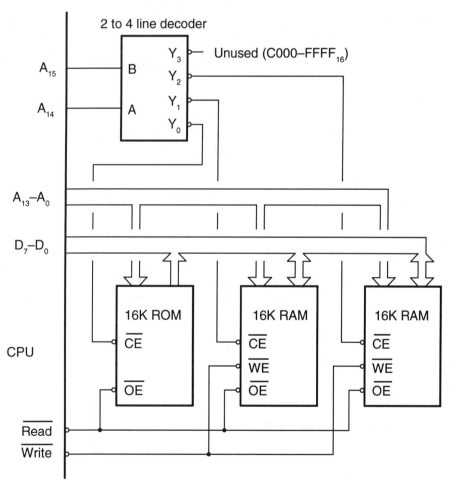

Fig. 1.29 Address decoding logic for the memory map of figure 1.11.

A commonly used, commercially available address decoder is the 74138 *3 to 8 line decoder* IC, which is part of the TTL logic family. Figure 1.30 gives details of this device (reproduced by courtesy of Texas Instruments).

This IC accepts a 3 bit address input and activates one of eight (active low) outputs, depending on the binary code applied to the inputs. Under normal circumstances the address inputs (A, B and C) would be connected to the most significant address bus lines (A_{15}, A_{14} and A_{13}), while the eight outputs would be used to provide memory chip enable signals. This basic circuit would be suitable for decoding up to eight memory devices, each having an eight kilobyte capacity.

The '138 also has three enable inputs, two of which are active low, while the other is active high. These inputs may be hard-wired to their respective logic levels, or may be connected to other signals in more complex address decoding circuits. (See the practical exercise on page 28 for some example circuits.) All three enable inputs must be at the correct logic levels for the decoder outputs to be enabled.

Practical Exercise   3 to 8 Line Decoder IC

Method

Connect the three enable inputs to the appropriate logic levels in order to allow the 74LS138 to function normally (G1 = logic 1, $\overline{G2A} = \overline{G2B} = $ logic 0).

Complete the following truth table, and hence comment on the operation of this device.

A B C	$\overline{0}$ $\overline{1}$ $\overline{2}$ $\overline{3}$ $\overline{4}$ $\overline{5}$ $\overline{6}$ $\overline{7}$
0 0 0	
0 0 1	
0 1 0	
0 1 1	
1 0 0	
1 0 1	
1 1 0	
1 1 1	

Confirm that the IC is disabled unless all three enable inputs are at the required logic levels.

Equipment required

74LS138 3 to 8 line decoder IC
Prototype board
5 volt d.c. power supply
Logic probe

74LS138 Schematic Layout

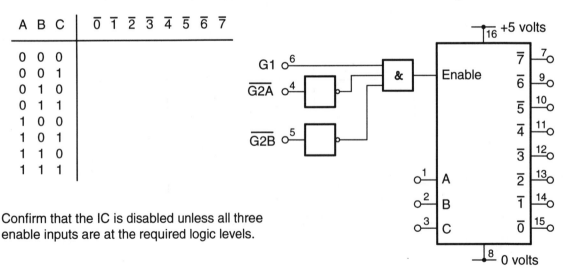

TYPES SN74LS138, SN74S138A, SN54LS138, SN54S138A
3-LINE TO 8-LINE DECODERS/DEMULTIPLEXERS

DECEMBER 1972 – REVISED APRIL 1985

- **Designed Specifically for High-Speed:**
  Memory Decoders
  Data Transmission Systems

- **3 Enable Inputs to Simplify Cascading and/or Data Reception**

- **Schottky-Clamped for High Performance**

- **Package Options Include Standard Plastic (N) and Ceramic (J) 300-mil Dual-In-Line Packages, Plastic Small Outline (D) and Ceramic Chip Carrier (FK) Package**

description

These Schottky-clamped TTL MSI circuits are designed to be used in high-performance memory decoding or data-routing applications requiring very short propagation delay times. In high-performance memory systems these decoders can be used to minimize the effects of system decoding. When employed with high-speed memories utilizing a fast enable circuit the delay times of these decoders and the enable time of the memory are usually less than the typical access time of the memory. This means that the effective system delay introduced by the Schottky-clamped system decoder is negligible.

The 'LS138 and 'S138A decode one of eight lines dependent on the conditions at the three binary select inputs and the three enable inputs. Two active-low and one active-high enable inputs reduce the need for external gates or inverters when expanding. A 24-line decoder can be implemented without external inverters and a 32-line decoder requires only one inverter. An enable input can be used as a data input for demultiplexing applications.

All of these decoder/demultiplexers feature fully buffered inputs, each of which represents only one normalized load to its driving circuit. All inputs are clamped with high-performance Schottky diodes to suppress line-ringing and to simplify system design.

The SN54LS138 and SN54S138A are characterized for operation over the full military temperature range of −55°C to 125°C. The SN74LS138 and SN74S138A are characterized for operation from 0°C to 70°C.

SN54LS138, SN54S138A ... J PACKAGE
SN74LS138, SN74S138A ... D OR N PACKAGE
(TOP VIEW)

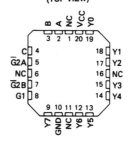

SN54LS138, SN54S138A ... FK PACKAGE
(TOP VIEW)

NC – No internal connection

logic symbols

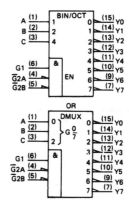

Pin numbers shown on logic notation are for D, J or N packages.

TEXAS INSTRUMENTS

3-297

Fig. 1.30 (a) 3 to 8 line decoder data sheets (courtesy of Texas Instruments).

TYPES SN74LS138, SN74S138A, SN54LS138, SN54S138A
3-LINE TO 8-LINE DECODERS/DEMULTIPLEXERS

logic diagram and function table

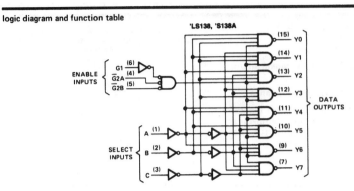

Pin numbers shown on logic notation are for D, J or N packages.

'LS138, 'S138A
FUNCTION TABLE

INPUTS					OUTPUTS							
ENABLE		SELECT										
G1	Ḡ2*	C	B	A	Y0	Y1	Y2	Y3	Y4	Y5	Y6	Y7
X	H	X	X	X	H	H	H	H	H	H	H	H
L	X	X	X	X	H	H	H	H	H	H	H	H
H	L	L	L	L	L	H	H	H	H	H	H	H
H	L	L	L	H	H	L	H	H	H	H	H	H
H	L	L	H	L	H	H	L	H	H	H	H	H
H	L	L	H	H	H	H	H	L	H	H	H	H
H	L	H	L	L	H	H	H	H	L	H	H	H
H	L	H	L	H	H	H	H	H	H	L	H	H
H	L	H	H	L	H	H	H	H'	H	H	L	H
H	L	H	H	H	H	H	H	H	H	H	H	L

*Ḡ2 · Ḡ2A · Ḡ2B
H = high level, L = low level, X = irrelevant

TYPES SN74LS138, SN74S138A, SN54LS138, SN54S138A
3-LINE TO 8-LINE DECODERS/DEMULTIPLEXERS

schematics of inputs and outputs

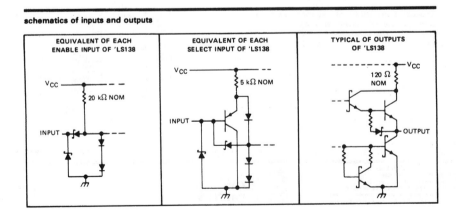

Fig. 1.30 (b) 3 to 8 line decoder data sheets (courtesy of Texas Instruments).

TYPES SN74LS138, SN54LS138
3-LINE TO 8-LINE DECODERS/DEMULTIPLEXERS

absolute maximum ratings over operating free-air temperature range (unless otherwise noted)

Supply voltage, V_{CC} (see Note 1) .... 7 V
Input voltage .... 7 V
Operating free-air temperature range: SN54LS138 .... −55°C to 125°C
　　　　　　　　　　　　　　　　　　SN74LS138 .... 0°C to 70°C
Storage temperature range .... −65°C to 150°C

NOTE 1: Voltage values are with respect to network ground terminal.

recommended operating conditions

		SN54LS138			SN74LS138			UNIT
		MIN	NOM	MAX	MIN	NOM	MAX	
V_{CC}	Supply voltage	4.5	5	5.5	4.75	5	5.25	V
V_{IH}	High-level input voltage	2			2			V
V_{IL}	Low-level input voltage			0.7			0.8	V
I_{OH}	High-level output current			− 0.4			− 0.4	mA
I_{OL}	Low-level output current			4			8	mA
T_A	Operating free-air temperature	− 55		125	0		70	°C

electrical characteristics over recommended operating free-air temperature range (unless otherwise noted)

PARAMETER	TEST CONDITIONS †		SN54LS138			SN74LS138			UNIT
			MIN	TYP‡	MAX	MIN	TYP‡	MAX	
V_{IK}	V_{CC} = MIN,　I_I = − 18 mA				− 1.5			− 1.5	V
V_{OH}	V_{CC} = MIN,　V_{IH} = 2 V,　V_{IL} = MAX, I_{OH} = − 0.4 mA		2.5	3.4		2.7	3.4		V
V_{OL}	V_{CC} = MIN,　V_{IH} = 2 V, V_{IL} = MAX	I_{OL} = 4 mA	0.25	0.4		0.25	0.4		V
		I_{OL} = 8 mA				0.35	0.5		
I_I	V_{CC} = MAX,　V_I = 7 V				0.1			0.1	mA
I_{IH}	V_{CC} = MAX,　V_I = 2.7 V				20			20	μA
I_{IL}	V_{CC} = MAX,　V_I = 0.4 V	Enable			− 0.4			− 0.4	mA
		A, B, C			− 0.2			− 0.2	
I_{OS} §	V_{CC} = MAX		− 20		− 100	− 20		− 100	mA
I_{CC}	V_{CC} = MAX,　Outputs enabled and open			6.3	10		6.3	10	mA

†For conditions shown as MIN or MAX, use the appropriate value specified under recommended operating conditions.
‡All typical values are at V_{CC} = 5 V, T_A = 25°C.
§Not more than one output should be shorted at a time.

switching characteristics, V_{CC} = 5 V, T_A = 25°C

PARAMETER¶	FROM (INPUT)	TO (OUTPUT)	LEVELS OF DELAY	TEST CONDITIONS	SN54LS138 SN74LS138			UNIT
					MIN	TYP	MAX	
t_{PLH}	Binary Select	Any	2	R_L = 2 kΩ,　C_L = 15 pF, See Note 2		11	20	ns
t_{PHL}						18	41	ns
t_{PLH}			3			21	27	ns
t_{PHL}						20	39	ns
t_{PLH}	Enable	Any	2			12	18	ns
t_{PHL}						20	32	ns
t_{PLH}			3			14	26	ns
t_{PHL}						13	38	ns

¶t_{PLH} = propagation delay time, low-to-high-level output; t_{PHL} = propagation delay time, high-to-low-level output.
NOTE 2: See General Information Section for load circuits and voltage waveforms.

TEXAS INSTRUMENTS

Fig. 1.30 (c) 3 to 8 line decoder data sheets (courtesy of Texas Instruments).

Practical Exercise

Address Decoding Circuits

Method

Connect each test circuit to the appropriate microprocessor address bus signals (using buffered signals).

Predict the address ranges which will be decoded by each output pin. Verify each prediction by writing a simple test program to read the content of a typical memory location in each address range. Run the test program and confirm (using the logic probe) that a pulse is produced on the correct address decoder output pin.

Equipment Required

74LS138 3 to 8 line decoder IC
Prototype board
5 volt d.c. power supply
Logic probe (with pulse detector facility)
Z80 or 6502 based microcomputer

Test Circuit 'A'

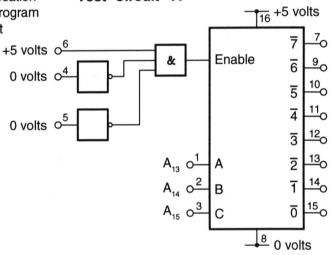

Truth Table 'A'

Active Output	Address Range
$\overline{0}$	
$\overline{1}$	
$\overline{2}$	
$\overline{3}$	
$\overline{4}$	
$\overline{5}$	
$\overline{6}$	
$\overline{7}$	

Truth Table 'B'

Active Output	Address Range
$\overline{0}$	
$\overline{1}$	
$\overline{2}$	
$\overline{3}$	
$\overline{4}$	
$\overline{5}$	
$\overline{6}$	
$\overline{7}$	

Test Circuit 'B'

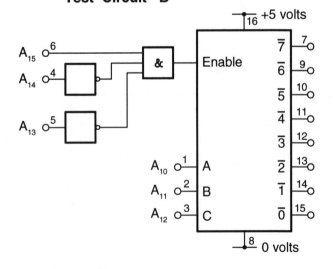

Ambiguous Decoding

When designing address decoding circuitry for a microcomputer, cost constraints may require that the number of components used is kept to the minimum. Such pressures may lead the designer to produce memory system designs which are not fully decoded, as shown in figure 1.31.

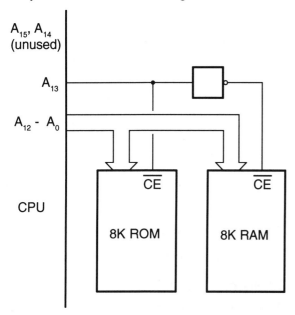

Fig. 1.31 Ambiguous address decoding.

In this case, an unused inverting gate from some other part of the computer has been used for the address decoding circuit, rather than use a 3 to 8 line decoder IC. The state of address line A_{13} is used to enable the two 8 kilobyte memory devices, leaving address lines A_{15} and A_{14} unused.

In principle, 48 kilobytes of memory space remains available for future memory expansion, but due to the ambiguous nature of the decoding circuitry, no future expansion is possible! The memory map is filled with repeating copies of the RAM and ROM, as shown in figure 1.32.

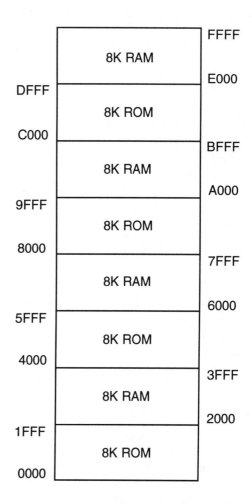

Fig. 1.32 Memory map caused by ambiguous address decoding circuitry.

In general, *ambiguous decoding* should be avoided as it reduces the possibilities for future microcomputer expansion. It is better to design memory systems with expansion in mind, rather than to prevent it!

Other Microcomputer Circuits

In addition to the memory, address decoding and buffer circuits considered previously, there are several other microcomputer circuits which are commonly encountered. These are briefly outlined here, to allow the interpretation of a complete microcomputer block diagram.

D.C. Power Supply

A fully stabilised d.c. power supply is a basic requirement. Most modern microcomputers require a single 5 volt supply, which must remain within certain limits to ensure correct operation. (As an example of this, TTL ICs require a supply which remains in the range 4.75 to 5.25 volts.)

The power supply may consist of transformer, rectifier, regulator and smoothing circuits, or may be of the more modern *switched mode* variety.

Microcomputers with low power consumption may be suitable for battery powered operation. CMOS components would normally be used in this case, due to their low power consumption and wide operating voltage range.

Input/output Devices

It is the input/output section which allows communication with peripheral devices, and the 'outside world' in general.

A wide variety of such devices may be encountered including the input/output ports and keyboard interfaces which were introduced in the first volume. Other input/output devices which will be discussed later include industry standard serial and parallel interfaces, conversion between analogue and digital signal types and multiplexed 7-segment displays. (More complex or specialised devices may also be encountered such as disc drive or display controllers which are beyond the scope of this text.)

In each case, communication between the microprocessor and the peripheral takes place across a purpose designed interface which involves connection to the microprocessor buses and address decoding circuitry.

Reset Circuit

Following the application of power (or pressing the *reset* button), the microprocessor may be forced to begin running ROM-based software, such as a *monitor*, *operating system* or *bootstrap loader*. This is achieved using an RC network and a switch, as shown in figure 1.33.

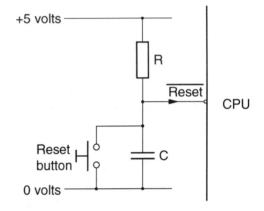

Fig. 1.33 Typical power-on reset circuit.

(The reset input is assumed to be active low here.)

Oscillator Circuit

The microprocessor clock waveform is generated by an oscillator. The frequency determining component is normally a *quartz crystal*, due to its high frequency stability. Figure 1.34 shows typical connections to the microprocessor.

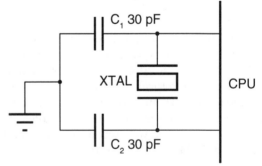

Fig. 1.34 Typical oscillator connections to the microprocessor (internal oscillator).

Figure 1.35 shows a complete microcomputer block diagram, containing all of the elements introduced in this chapter.

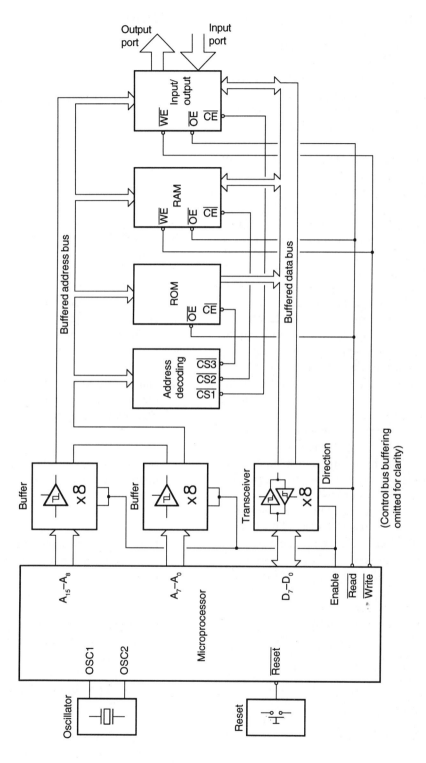

Fig. 1.35 Complete microcomputer block diagram.

Summary

The main points covered in this chapter are:

- The *central processing unit (CPU)* of a *micro-computer* is called a *microprocessor*.
- The microprocessor contains *registers* which may refer to data or to the address in memory of an item of data. Many registers (such as the *flags register*, *program counter* or *stack pointer*) have special purposes.
- Communication between the microprocessor and other components is via the address, data and control buses. Data travels to or from the microprocessor across the *data bus* from a single address, selected by the *address bus*. The microprocessor controls and synchronises all data transfers using specific *control bus* signals.
- The significance and timing of relevant microprocessor signals may be understood by referring to the appropriate *timing diagram*. Timing diagrams provide other useful information to the system designer, such as maximum *access times* for memory or input/output devices.
- Any device which is connected as an output to the data bus, must be capable of placing its outputs in a high impedance state, under microprocessor control (*tri-state* outputs).
- In complex computer systems, microprocessor output currents may become excessive, necessitating the use of *buffers*. Unidirectional buffers are suitable for the address and control buses, while bi-directional *transceiver* circuits may be needed for the data bus. Some buffers use *Schmitt trigger* inputs to provide increased *noise immunity*.
- Contention between memory or input/output devices is avoided by allocating non-overlapping memory ranges to each device. *Address decoding* logic is used to ensure that only one component receives a valid chip enable signal at any time. (*Ambiguous decoding* should be avoided where possible, as it tends to limit future system expansion.)

Problems

1 Define the following terms.

a) Microprocessor
b) Microcomputer
c) CPU
d) Register

2 State the purpose and characteristics of the three microprocessor buses.

3 Identify the main function of the following registers.

a) Accumulator
b) Program counter
c) Flags register
d) Stack pointer

4 Name, and briefly describe, the five machine cycles which are supported by the Z80 microprocessor.

5 Name, and briefly describe, the three machine cycles which are supported by the 6502 microprocessor.

6 Explain why tri-state outputs must be fitted to all devices capable of controlling the data bus.

7 A microprocessor output pin is capable of supplying an output current of 2 mA. This pin is connected to five inputs, each with a maximum input current of 0.5 mA. Should this signal be buffered and why?

8 Explain how address decoding is used to prevent bus contention, and its relationship to the system memory map.

2   Memory Devices

Aims

When you have completed this chapter you should be able to:

1 Appreciate the range of transistor types that is available, and their basic operation.
2 Understand the construction of commonly available logic families, such as TTL, CMOS and ECL.
3 Explain the operation of available types of RAM and ROM.
4 Select and use commercially available memory devices.

Introduction

In order to understand the operation of the various circuits presented in this chapter, a sound knowledge of transistor types and their operation is required. For this reason, available types of bipolar and unipolar transistors are briefly outlined here, together with commonly encountered applications.

Those readers who require a more detailed discussion of transistor operation, taking into account semiconductor theory, are referred to *Electronics: Practical Applications and Design* by John C. Morris and published by Edward Arnold. This text provides an easy to understand (non-mathematical) explanation of semiconductor based devices.

Transistor Types and Operation

The basic operation of *BJT* (*bipolar junction transistor*), *MOSFET* (*metal oxide semiconductor field effect transistor*) and *JFET* (*junction FET*) transistors is considered in the following sections, with particular reference to circuits found in commonly available logic families.

Bipolar Junction Transistor

This type of transistor is formed from three layers of silicon, and is available in *NPN* and *PNP* forms, as shown in figure 2.1.

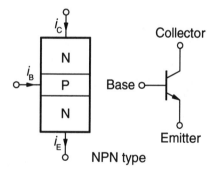

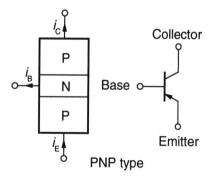

Fig. 2.1 NPN and PNP transistor types.

The bipolar transistor operates as a current amplifier, with a small current flowing at the base terminal controlling a much larger current at the collector terminal. The *current gain* (h_{FE}) of the transistor is given by the ratio of collector to base current.

Operation of NPN and PNP transistors is similar, except for the opposite directions of current at each terminal, which also results in different bias voltage potentials. To cause base current to flow in an NPN transistor, it is necessary to make the base terminal more positive than the emitter by approximately 0.7 volts. The same result is achieved with a PNP transistor by making the base more negative than the emitter by 0.7 volts. In each case the voltage against current characteristics of the base-emitter junction are those of a forward biased diode. Figure 2.2 shows the connection of an NPN transistor, operating as an on/off switch, while figure 2.3 gives the same information for a PNP transistor.

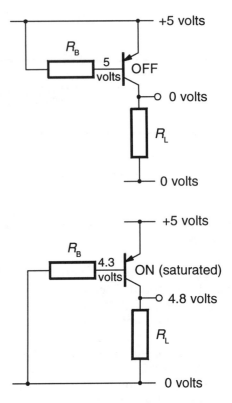

Fig. 2.3 PNP transistor used as a switch.

In each case it has been assumed that a saturated transistor will have a collector to emitter voltage of approximately 0.2 volts.

Another commonly encountered circuit is the *emitter follower*, in which the voltage on the emitter terminal 'follows' the voltage applied to the base terminal. This circuit is shown in figure 2.4.

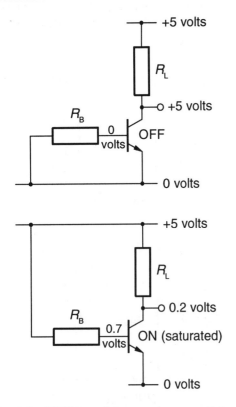

Fig. 2.2 NPN transistor used as a switch.

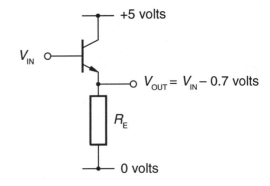

Fig. 2.4 Emitter follower circuit.

In the emitter follower circuit, the current flowing out of the emitter terminal flows through the resistor, thus developing a potential on the output. The current rises until $V_{BE} = 0.7$ volts and if V_{IN} changes, the emitter current alters accordingly. As will be seen later, this circuit is used in *ECL* (*emitter coupled logic*) circuitry.

The *phase splitter*, which is shown in figure 2.5, is a slight modification on the emitter follower.

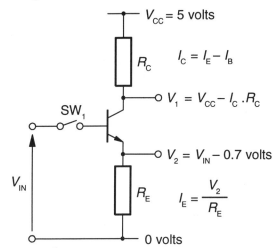

Fig. 2.5 Phase splitter circuit.

The base current (and base voltage) to the phase splitter is normally supplied by another section of the circuit. If the base current is absent, the transistor is off and the collector terminal will be at the positive supply voltage, while the emitter will be at 0 volts.

When a base current is present, the lower half of the circuit acts as an emitter follower. The current flowing through R_E (I_E) is controlled by V_{IN}. If V_{IN} increases then V_2 and I_E increase accordingly.

An increase in I_E causes a corresponding increase in I_C. (I_E and I_C are roughly equal if h_{FE} is large.) This increase in I_C causes a decrease in the voltage at the collector terminal (V_1), due to the volt-drop across R_C. Thus the presence of a base current causes V_1 to decrease and V_2 to increase. The phase splitter gets its name from the fact that the two output signals are out of phase!

The *long-tailed pair* is also commonly encountered (in ECL circuits), and is shown in figure 2.6.

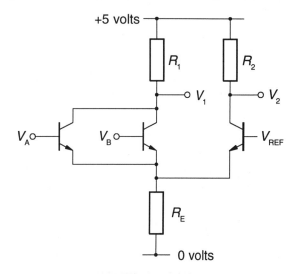

Fig. 2.6 The long-tailed pair circuit.

This circuit has two inputs (V_A and V_B) and two outputs (V_1 and V_2). The base of the right-hand transistor is also connected to a fixed reference voltage.

The long-tailed pair acts as a high speed logic circuit with both OR and NOR outputs! Circuit operation is based on causing current to flow through either R_1 or R_2, depending on the voltages applied to inputs A and B. If current flows through R_1, then V_1 will be at a 'low' logic level, while the other output will be 'high'. Since current never flows through R_1 and R_2 simultaneously, the outputs are always complementary (at opposite logic levels).

The lower half of the circuit acts as an emitter follower, with three transistors sharing a single emitter resistor. If V_A and V_B are significantly lower than the reference voltage, then current will flow through the right-hand branch of the circuit (through R_2), giving a low logic level on output 2. If either V_A or V_B (or both) are raised significantly above V_{REF}, then current flows though R_1, causing V_1 to be at logic 0.

It can be seen from this description that outputs V_1 and V_2 provide NOR and OR functions respectively on inputs A and B.

The MOSFET Transistor

The BJT (bipolar junction transistor) seen previously was a current controlled device, whose operation was based on the flow of both electrons and 'holes' (hence the term *bipolar*). The *field effect transistor* or *FET* is a voltage controlled device, operating due to the flow of either electrons or holes. The FET is therefore called a unipolar transistor.

Two basic types of MOSFET are available, These are the *n-channel FET*, in which current flow is due to electrons and the *p-channel FET* which conducts through the flow of holes. Figure 2.7 shows the construction of an n-channel MOSFET.

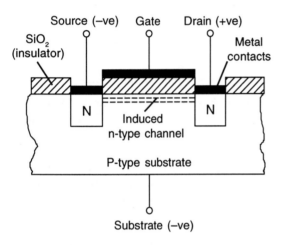

Fig. 2.7 N-channel MOSFET construction.

Operation is based on the creation of an n-type conducting channel between the source and drain terminals. The width (and hence the resistance) of the channel may be varied by attracting or repelling electrons from the area immediately below the gate contact. Electrons may be attracted by making the gate more positive (opposite charges attract), or may be repelled by making the gate more negative. If electrons are attracted, this makes the channel wider, reducing the source to drain resistance. Conversely the repulsion of electrons reduces the width of the channel, thus increasing the channel resistance.

Electrical isolation between different devices on the same silicon wafer is achieved by making the p-type substrate more negative than the n-type transistors. The substrate is connected to the negative supply rail, so each transistor is surrounded by a non-conducting *depletion region*, similar to a reverse biased diode.

Both *enhancement* and *depletion* type n-channel MOSFETs are available, as shown in figure 2.8.

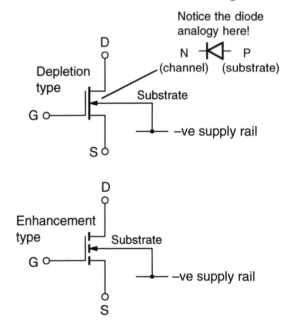

Fig. 2.8 Enhancement and depletion type n-channel MOSFETs.

In the depletion type of FET a conducting channel exists between the drain and source terminals, even if the gate is left unconnected. The channel in an n-channel depletion type FET may be eliminated by making the gate terminal more negative than the source (repelling electrons from the channel).

In an enhancement type FET, no channel exists in the absence of a gate voltage. To cause conduction between drain and source, charge carriers must be attracted to the area under the gate terminal, thus forming a conducting channel. So for an n-channel enhancement type FET the gate must be made more positive than the source terminal (attracting electrons) to allow conduction.

If you examine the circuit symbols for n-channel MOSFETs, it can be seen that the channel in an enhancement type FET is shown as a dotted line (indicating that the channel does not exist in the absence of a suitable gate potential). In the depletion type FET the channel is indicated as a solid line. The electrical insulation of the gate terminal, which is due to a thin SiO_2 layer, is shown as a small gap between the gate and the channel. (Another name for this type of transistor is the *IGFET* or *insulated gate FET*.) Notice also that the arrow on the substrate terminal is based on a diode analogy, and indicates both the channel and substrate types. Compare these circuit symbols with those for p-channel enhancement and depletion type MOSFETs, which are shown in figure 2.9.

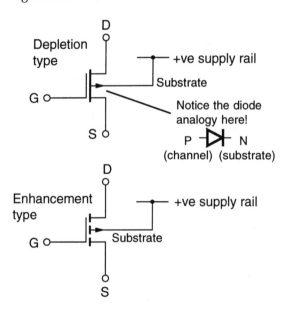

Fig. 2.9 Enhancement and depletion type p-channel MOSFETS.

Conduction in the p-channel MOSFET is due to the flow of holes, which behave as positive charge carriers. For this reason conduction occurs in the p-channel enhancement type transistor when the gate is made more negative than the source (opposite charges attract). Conversely a p-channel depletion type MOSFET will conduct until the gate is made significantly more positive than the source.

Table 2.1 summarises the switching characteristics of n-channel and p-channel enhancement and depletion type MOSFETs.

Type	V_{GS} +ve	V_{GS} 0	V_{GS} −ve
n-channel enhancement	ON	OFF	OFF
n-channel depletion	ON	ON	OFF
p-channel enhancement	OFF	OFF	ON
p-channel depletion	OFF	ON	ON

Table 2.1 MOSFET switching characteristics.

Another type of FET which is sometimes encountered is the *JFET* or *junction FET* (also known as the *JUGFET* or *junction gate FET*).

As the name suggests, the gate of the JFET is formed from a pn junction, which must be reverse biased in order to maintain electrical isolation between the gate and the channel. Figure 2.10 shows the construction of an n-channel JFET.

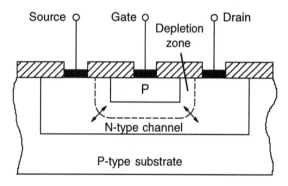

Fig. 2.10 N-Channel JFET construction.

As the gate is made more negative (more positive in a p-channel device), the depletion zone widens, thus increasing the resistance of the channel. Figure 2.11 shows standard JFET symbols.

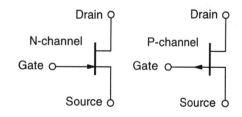

Fig. 2.11 N and p-channel JFET symbols.

Common Logic Families

The following section describes the major logic families, each of which is commonly available. Emphasis is placed on understanding the internal operation of typical circuits, in order to gain an appreciation of the areas of application of each logic family.

TTL

Transistor transistor logic (TTL) was first announced in 1964 by Texas Instruments. Since then, TTL has been one of the most widely used logic families and several improved versions are now available. Figure 2.12 illustrates a basic three input NAND gate.

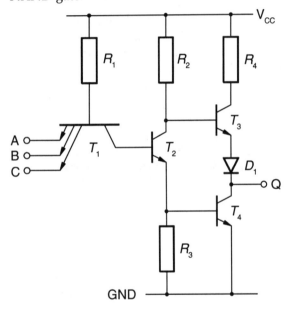

Fig. 2.12 TTL three input NAND circuit.

The circuit has three inputs and a single output and performs a logical NAND function (the output is low if all inputs are high). Circuit operation may best be understood by dividing the circuit into three main sections. From left to right these are,

- multi-emitter transistor,
- phase splitter,
- totem pole output circuit.

The multi-emitter NPN transistor (T_1) may be thought of as several individual diodes, as shown in figure 2.13.

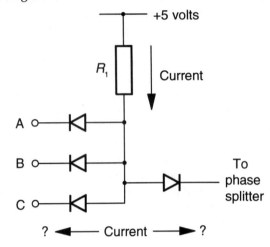

Fig. 2.13 Equivalent circuit of T_1.

Current which flows through R_1 from the positive supply rail is directed either to the right or to the left, depending on the logic levels applied to the three inputs. If all inputs are at logic 1 then each input diode is reverse biased and the current flows to the right (to the phase splitter). If one or more of the inputs is at a low logic level, then the current flows to the left (out of the input!).

The second stage is a phase splitter, which is turned on or off, based on the presence or absence of a base current to transistor T_2. If T_2 is off, a high impedance exists between the collector and emitter terminals, causing these terminals to be pulled toward the positive and negative supply rails respectively. When T_2 is on, the potential at the collector of T_2 falls, while that of the emitter rises.

A low impedance output drive is provided by the totem pole circuit. Transistor T_3 is turned on to produce a logic 1, while a logic 0 occurs if T_4 is made to conduct. (The phase splitter ensures that T_3 and T_4 never conduct simultaneously). If the phase splitter is off then T_3 receives base current via R_2, producing a logic 1 output. At the same time, the base of T_4 is at 0 volts, so T_4 is off. When the phase splitter is on, T_3 is off (no base current), while T_4 conducts, producing a logic 0 output.

In order to understand the function of D_1 it is necessary to consider the potentials at each point in the circuit. Figure 2.14 illustrates typical voltages for a logic 0 output.

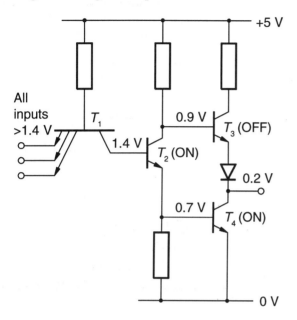

Fig. 2.14 TTL (logic 0) voltage levels.

Examining the above diagram, it can be seen that with T_4 and T_2 on, a potential of 0.7 volts appears between the base of T_3 and the collector of T_4 (the output). For a logic 0 output, transistor T_3 should be off, while T_4 is on. This is guaranteed by the presence of D_1, which provides a second pn junction in series with the base-emitter of T_3. (For T_3 to conduct, its base must be more positive than the output by approximately 1.4 volts.)

Several variations on the basic TTL circuit are commonly encountered. These include,

- tri-state outputs,
- open collector outputs,
- Schottky transistors.

In the tri-state output circuit, additional logic is added so that both output transistors (T_3 and T_4) can be simultaneously disabled, under the control of a tri-state input. This causes the output terminal to be electrically isolated, which is commonly required in bus-based systems.

(Recall that the need for tri-state outputs for use with circuits capable of outputting data onto the data bus was explained in chapter one.)

Certain TTL ICs are available with *open collector* output circuits, as shown in figure 2.15.

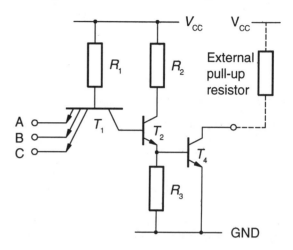

Fig. 2.15 TTL NAND gate with an open collector output.

In this case the totem pole output stage is replaced with a transistor whose collector is linked directly to the output terminal. The output of such a gate is incapable of asserting a logic 1 level directly, without the help of an external pull-up resistor. The main advantage of open collector outputs is that several such outputs may be wired together, sharing a single pull-up resistor, which produces a 'wired' logic function. This is normally called a *wired-AND* function for positive logic (active high) signals, or *wired-OR* in negative logic (active low) applications. (An example using wired-OR connection of outputs is considered in the next chapter.)

A final modification to the basic TTL gate is the use of *Schottky transistors*, which allow higher operating speeds or lower power consumption to be achieved. This is based on the connection of a diode with a low forward voltage between the base and the collector of each transistor, as shown in figure 2.16 overleaf.

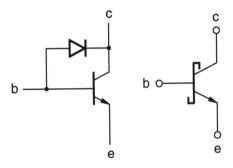

Fig. 2.16 Schottky transistor and symbol.

Logic families such as TTL, which rely on transistor saturation for their operation, are limited in operating speed by the minimum time taken for each transistor to turn on and then off again. Transistor saturation causes excess charge to be stored in the base region which must be discharged before the transistor can turn off. It is this unwanted capacitance which limits the maximum operating frequency.

The addition of a normally reverse biased diode between the base and collector terminals has no effect during normal transistor operation (where the collector is more positive than the base). When the transistor enters saturation, the diode becomes forward biased, causing excess base current to be harmlessly diverted away from the base terminal, hence limiting the degree of saturation.

The Schottky diode, which has a low forward voltage, is formed from a metal to semiconductor junction. In practice, this is constructed by extending the metal connection to the base terminal slightly over the collector region.

NMOS, PMOS and CMOS

A number of closely related logic families are available, whose operation is based on particular types of MOSFET. These include,

- *NMOS*   *n-channel MOS* transistors,
- *PMOS*   *p-channel MOS* transistors,
- *CMOS*   *complementary MOS*
          (combining both n-channel and
          p-channel devices in a single IC)

In general, CMOS devices are more expensive due to the need to fabricate both n-channel and p-channel devices on a single silicon wafer. However CMOS devices have several advantages over NMOS and PMOS, including lower consumption and a symmetrical output drive, as will be seen.

Figure 2.17 shows an NMOS NOR gate.

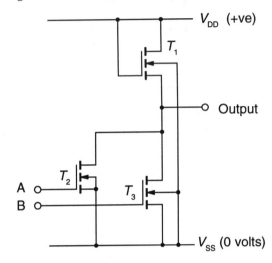

Fig. 2.17 NMOS NOR gate.

The gate of TR_1 is connected directly to the positive supply, causing a constant resistance between the drain and source terminals. TR_1 acts as a pull-up resistor, causing the output to be at logic 1 when TR_2 and TR_3 are both off.

If either TR_2 or TR_3 turns on, the output becomes logic 0. (This may be achieved by making the gate more positive than the source in an n-channel enhancement type transistor.)

As can be seen the circuit produces a NOR logic function. However, the output drive is non symmetrical, due to the passive generation of the logic 1 output level.

One disadvantage of MOS circuits is the very high input impedance, caused by the use of an insulated gate connection. This makes input terminals susceptible to damage by static electricity. In practice clamping diodes are used to protect input circuits from static damage, but special handling precautions must still be observed with this type of IC.

A CMOS inverter is shown in figure 2.18.

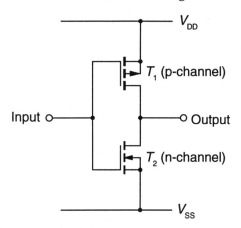

Fig. 2.18 CMOS inverter.

The inverter is formed by connecting two transistors in parallel. The p-channel device is on when the input is connected to the negative supply rail, while a logic 1 input is required for the n-channel device to conduct. Thus if the input is high, T_2 conducts and the output is logic 0. If the input is low, T_1 conducts, giving a logic 1 output.

Figure 2.19 shows a CMOS NOR gate.

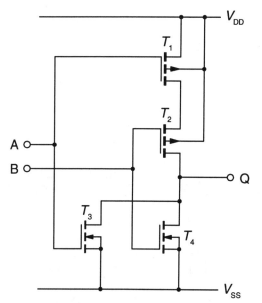

Fig. 2.19 CMOS NOR gate.

For a logic 1 output, both T_1 and T_2 must be on (due to their series connection), while a logic 0 is generated if either T_3 or T_4 (or both) conducts.

If both inputs are at logic 0 then T_3 are T_4 are both off (n-channel devices), while T_1 and T_2 (p-channel transistors) are both on. Thus the output is at logic 1. If one or more of the inputs is at logic 1, then at least one of the transistors T_3 or T_4 will conduct, while either T_3 or T_4 (or both) will be off, giving a logic 0 output.

It is left as an exercise for the reader to devise a CMOS circuit which produces a NAND logic function. (The circuit of figure 2.19 is a good starting point.)

Notice that in CMOS circuits, both logic levels are actively generated, allowing a low impedance output drive at each logic level. This means that CMOS devices have very low power consumption. In fact, the main use of energy is due to the charging and discharging of parasitic capacitances which occurs when the transistors are turned on and off. For this reason, the power consumption of these circuits is very low for d.c., but rises almost linearly as the operating frequency rises.

As well as having low power consumption, CMOS ICs may be operated from a wide range of supply voltages (3 to 18 volts typically), making them very suitable for battery powered operation. A logic 0 is defined as being between 0–30% of the supply voltage while a logic 1 is greater than 70% of the supply.

CMOS devices also have improved noise immunity, compared with TTL, as their operation is not based on the forward biasing of pn junctions. When operated from a 5 volt supply the noise immunity is approximately 1 volt (compared with 0.4 volts for TTL).

The presence of unwanted capacitances, which was mentioned above, also limits the maximum operating frequency of CMOS, which is lower than that of TTL. Improved CMOS logic families have recently been developed, which are competitive with TTL. In fact, pin compatible replacements for many TTL circuits are now available which are based internally on CMOS logic! The 74HC logic family is pin compatible with TTL and uses CMOS logic levels, while 74HCT devices allow direct plug-in replacement of TTL ICs, due to their use of TTL logic levels. (Recall that the main characteristics of TTL and CMOS based logic families were illustrated in table 1.6).

ECL

Emitter coupled logic (ECL) offers higher operating speeds than either TTL or CMOS, making this logic family suitable for applications where high speed is essential. Unlike TTL (which is also based internally on bipolar transistors), ECL transistors are not allowed to enter saturation, resulting in higher switching speeds. Switching times of about 1 nanosecond (1×10^{-9} seconds) are possible using ECL logic. Figure 2.20 shows a typical ECL gate with complementary (OR/NOR) outputs.

The first thing to notice is that ECL is historically based on a negative supply voltage, with the most positive supply terminal being connected to the zero volt rail. Logic levels for logic 0 and logic 1 are defined as –1.6 and –0.8 volts respectively, so logic 1 is still the most positive of the two levels. (The reason for the small gap between the logic levels, which is actually one of ECL's disadvantages, will be seen in a moment.)

Both inputs are applied to a long-tailed pair circuit, with the reference voltage chosen to be midway between the two logic levels at –1.2 volts.

As seen previously (figure 2.6 and the associated text), current flows through either R_1 or R_2 depending on the voltages applied to inputs A and B. For complete current switching the input voltages must differ from the reference voltage by at least 0.4 volts, resulting in the ECL logic levels seen earlier. The current consumed by the circuit is relatively constant at all times, preventing the current spikes caused by changes in logic level, which are a problem with other logic families.

Connections are taken from R_1 and R_2, which provide the complementary (OR/NOR) outputs via a pair of emitter followers. These circuits produce a potential of 0.7 volts across each base emitter junction, which ensures that correct ECL output logic levels are provided. With a logic 1 output (–0.8 volts), approximately 0.1 volts is developed across R_1 or R_2 (by the base current), while the remaining 0.7 volts is produced by the emitter follower. External connection of emitter resistors R_4 and R_5 also allows wired-OR connection of outputs.

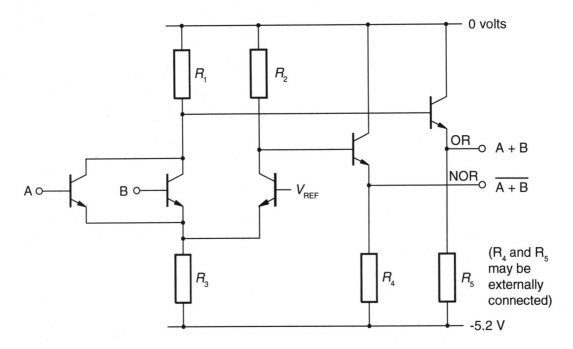

Fig. 2.20 Two input ECL OR/NOR gate.

Available Types of Memory

As was mentioned in chapter one, a microcomputer is normally fitted with both RAM and ROM. The following section considers commercially available types of memory while examining their methods of construction and internal operation. Figure 2.21 illustrates commonly available memory types.

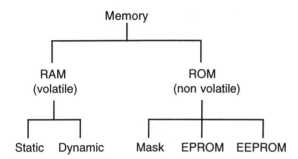

Fig. 2.21 Commonly available memory types.

Two types of RAM are normally encountered. Static RAM uses a bistable circuit as the basic storage element (each bistable holding one bit of information), while dynamic RAM uses the presence or absence of charge on a capacitor to represent the stored data. Both types of memory are said to be *volatile*, meaning that the information is lost when power is removed.

With dynamic RAM, it is also necessary periodically to *refresh* the memory to prevent the loss of stored data. Due to the tendency of stored charges to leak away, the charge on each capacitor must be 'topped-up' every few milliseconds. This requirement adds to the complexity of memory systems based on dynamic RAM, although this is compensated for by the simpler internal construction of dynamic RAM. In general, dynamic RAM is available in higher memory capacities and with lower power consumption than static RAM.

Several types of ROM are commonly available. In *mask programmed ROM*, the data content is defined during the final stage of manufacture, when a user-defined mask is overlaid onto the ROM. This method of production is only economical for large volume production, where the ROM content is fixed. In such applications, mask ROM is the cheapest form of non-volatile memory.

Smaller volume production is possible using *EPROMs* (*electrically programmable* ROM), which may be programmed using a special programming unit. These devices may later be erased by exposure to ultra-violet light, and may be re-used several times. EPROMS are more expensive than mask programmed ROM, but allow non-volatile memory to be programmed in small volume production situations. Their ability to be reprogrammed makes EPROMS ideal in prototype development situations.

Even greater flexibility is achieved using *EEPROMs* (*electrically programmable and erasable* ROM). EEPROMS may be programmed and erased electrically, and do not require a special programmer or unusual programming voltages. A typical EEPROM may be reprogrammed 100,000 times during its lifetime, without once being removed from its socket. This makes EEPROMS ideal for applications where non-volatile data storage is required, with the capacity for periodic alteration of data. An example of this would be a *point of sale terminal*, where the price of particular items may, from time to time, be altered. EEPROMS are more expensive than EPROMS and mask programmed ROMs but have the advantage of being re-programmable without the need for any additional equipment.

The next section considers the internal construction of memory devices in general, before examining the detailed circuit operation of particular types of RAM and ROM.

Internal Memory Organisation

The memory map of a microcomputer defines areas of memory which are either RAM or ROM. *Address decoding* circuitry ensures that the correct memory device is enabled at any time, and that all other devices are disabled. This address decoding circuitry is external to the memory.

When an address is applied to the address inputs of a memory device, this causes a single memory location within the memory to be selected, while all other locations are inactive. Once again, this is achieved using address decoding circuitry. In this case the address decoding circuitry is internal to the memory, as shown in figure 2.22.

The internal address decoding circuitry is normally arranged as a matrix, with each memory location connected to row and column decoder circuits. (This arrangement is simpler than a linear decoding system and offers faster memory access times.) For a memory location to be selected, the row and column select signals must both be active, which is equivalent to a logical AND function.

Each memory cell is connected to an internal data bus, which allows the content of the selected memory location to be read or (in the case of RAM) altered. This is illustrated by figure 2.23.

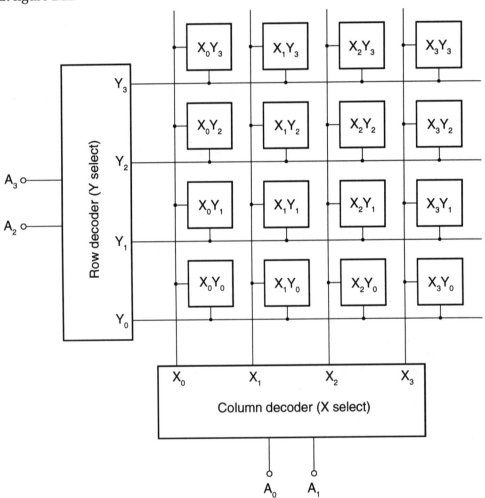

Fig. 2.22 Internal memory address decoding.

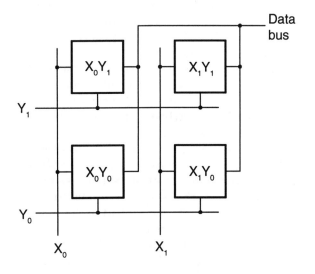

Fig. 2.23 Internal data bus connections.

The selected memory location (whose row and column inputs are both active) is able to read or write data via the data bus, while all other locations hold their outputs in a high impedance state.

Memory devices are available with various data widths, ranging from a single bit to 16 bits or more. In cases where the memory data width is less than that of the data bus, it is common practice to connect several memory devices in parallel. Each device then supplies part of the data required by the microprocessor. Figure 2.24 shows several possible arrangements.

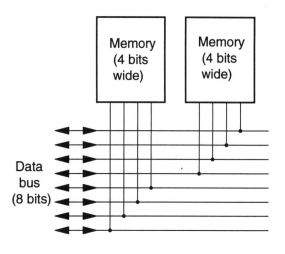

Fig. 2.24(b) 8 bit data bus with 4 bit memory.

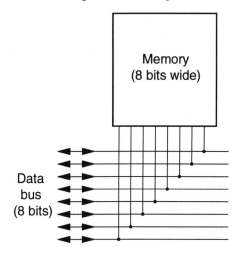

Fig. 2.24(a) 8 bit data bus with 8 bit memory.

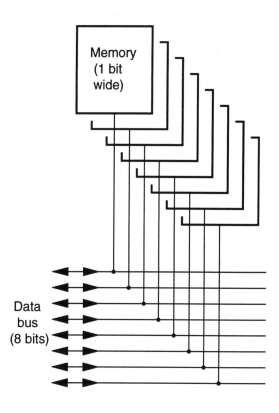

Fig. 2.24(c) 8 bit data bus with 1 bit memory.

Static RAM

As mentioned earlier, static RAM operation is based on the bistable, which is a 1 bit storage element. Figure 2.25 shows a simple NMOS bistable, which is formed from transistors T_1, T_2, T_3 and T_4.

The gates of T_1 and T_2 are both connected directly to the positive supply rail, causing them to act as load resistors for transistors T_3 and T_4. As the name *bistable* suggests, the circuit can exist in two stable states, with either T_3 or T_4 conducting, but never both. The state of the bistable may be altered by stopping or diverting the current which normally flows through the conducting transistor. The drain is then pulled-up to logic 1 by the load resistor. Due to the cross coupling of gate connections, this causes the second transistor to conduct, which in turn holds the first transistor in the 'off' state. The circuit remains in this condition until new data is stored, or the power supply is removed.

The drain terminals of T_3 and T_4 provide complementary data outputs, which are connected to the *data* and $\overline{data}$ lines in the figure, via two pairs of transistors. Transistors T_5, T_6, T_7 and T_8 are used to provide the output enable circuitry, and ensure that only the selected memory cell is connected to the data lines. Thus T_5 and T_6 are enabled by the row select decoder, while T_7 and T_8 are controlled by the column select circuitry.

Once a single memory location has been selected, its content may be determined by placing the data lines into a high impedance state and passively sensing the logic levels produced by the bistable. Conversely, the content of the bistable may be changed by actively driving the data lines and forcing the bistable into the desired state.

Although the example shown here uses NMOS technology, CMOS static RAM is also available. The lower power consumption of CMOS devices (particularly in *stand-by* mode) makes them suitable for battery-backed, non-volatile memory.

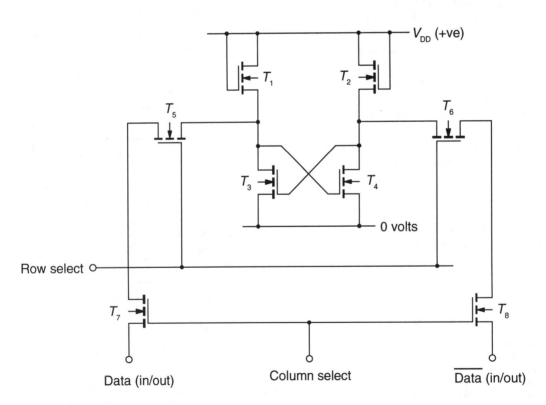

Fig. 2.25 NMOS bistable circuit.

Practical Exercise   Static RAM

Introduction

This exercise demonstrates the operation of a bistable, which is the basic storage element used by static RAM. The use of a battery to preserve memory content following the disconnection of the power supply is also shown.

Method

Connect the circuit shown, using a suitable prototype board.

a) By applying pulses to the two push to make switches, verify that the content of the memory may be changed, and that the circuit 'remembers' the last applied input.

b) With the battery connected, confirm that the memory is non-volatile (by turning the power supply off for a few seconds).

Equipment required

Two BC108 transistors
Three 10 k resistors
Two 470 Ω resistors
Two push to make switches
SPST switch
Two LEDs
Two 1N4001 diodes
9 volt battery
9 volt power supply
Prototyping board

Results

SW2	SW3	LED 1	LED 2
Pulse			
	Pulse		
Pulse			

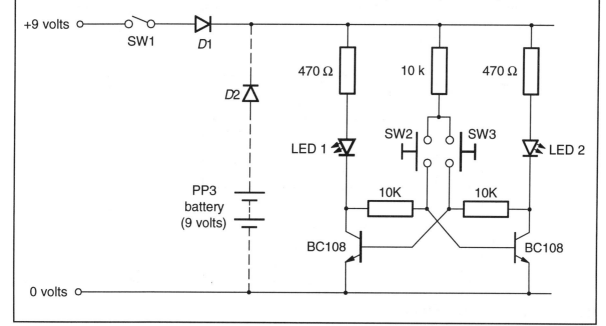

Dynamic RAM

Dynamic RAM uses the gate to channel capacitance of a FET to store binary information, as shown in figure 2.26.

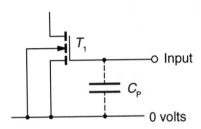

Fig. 2.26   Gate to channel capacitance.

Consider an example where a logic 1 input is momentarily applied to the above circuit (causing T_1 to switch on), after which the input line is placed in a high impedance state. This momentary input causes the parasitic capacitance C_P to be charged, which in turn holds T_1 in the 'on' state after the removal of the input. The circuit exhibits the property of *memory*.

Due to leakage of the stored charge, the transistor will eventually turn off, resulting in the loss of the stored data. For this reason the content of dynamic RAM must be regularly *refreshed*, which involves the recharging of any previously charged capacitors in the memory array. Figure 2.27 shows a practical dynamic RAM cell, capable of storing a single bit of information.

Once again, a bistable is used to store the binary information, in this case using only two transistors T_1 and T_2.

To store a single bit of information, the required data is applied to the two data lines and the address is selected by strobing the address select line (which enables T_3 and T_4). The complementary logic levels on the two data lines cause one of the bistable transistors to be switched on, while the other is turned off. When T_3 and T_4 are disabled, the parasitic capacitances C_{P1} and C_{P2} hold the bistable in the required state.

As for static RAM, the process of reading the data content involves placing the data lines in a high impedance state and pulsing the address select line. In the case of dynamic RAM, the data content of each location must also be refreshed every few milliseconds to prevent loss of data. This is achieved with a *sense amplifier* circuit, which basically consists of a *comparator* and a *voltage reference*.

One input of the comparator is connected to a reference voltage, which is at 50% of the supply. During a data read the comparator produces a logic 0 or a logic 1 by comparing the content of the bistable with the reference voltage. The output of the sense amplifier is connected to the data output of the memory and is also used to recharge the bistable capacitances to their optimum values.

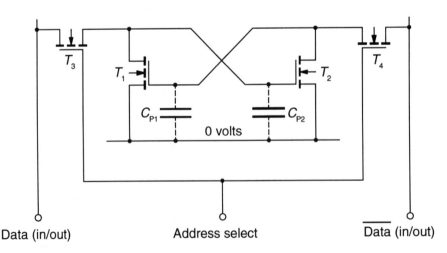

Fig. 2.27   Practical dynamic RAM cell.

Sense amplifier circuits are normally connected to each column in the memory array, and selecting a particular row causes each location in that row to be refreshed, as shown in figure 2.28.

As can be seen, selecting a particular row causes each output transistor in that row to be enabled. This in turn connects the internal voltage level to the sense amplifier for that column, which performs the refresh operation.

In practice, each row of the memory array must be read every few milliseconds to refresh the entire dynamic RAM. This is performed by cycling through each row address in a repeating sequence, which occurs under the control of the dynamic RAM refresh control circuitry.

Many modern dynamic RAMs have on-chip refresh control, which greatly simplifies the system design. Some microprocessors (such as the Z80) also have built-in support for dynamic RAM, as was mentioned in the first volume.

The added complexity of dynamic RAM, due to its need for refreshing is compensated for by the fact that larger memory capacities are available, compared with static RAM. This is due to the simpler construction of the dynamic RAM cell, which allows a greater number of memory arrays to be fabricated on a silicon wafer. The physical size of dynamic RAM ICs is also less than that of static RAM due to the common use of *multiplexed address inputs*. In this system the row and column inputs are applied to the same address inputs in sequence and are latched under the control of row and column *address strobe* signals. This results in a 50% reduction in the number of external address pins, compared with a static RAM of the same capacity. This reduction in size is important in the crowded printed circuit boards of modern computers!

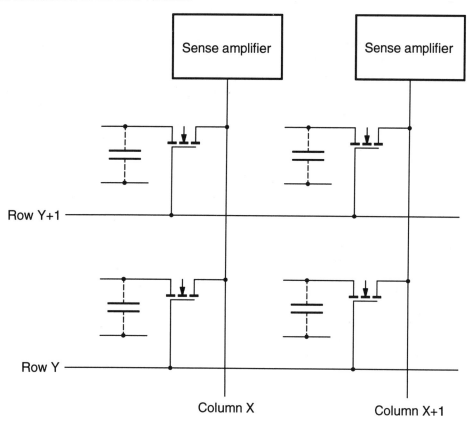

Fig. 2.28 Dynamic RAM refresh circuitry.

Practical Exercise   Dynamic RAM

Introduction

This exercise demonstrates the storage of a
single bit of information using a capacitor,
and the need for periodic data refreshing with
a sense amplifier circuit.

Method

Connect the circuit shown.

a) Momentarily connect the capacitor to the
   positive supply, noting the gradual
   leakage of stored charge using the digital
   voltmeter.

b) Observe the output of the sense amplifier
   circuit and use the comparator output to
   periodically recharge the capacitor.

Equipment required

Comparator (LM311 or equivalent)
470 k resistor
100 µF capacitor
Two 10 k resistors
Digital voltmeter
Prototyping board

Notes

Any suitable values may be substituted for
the 470 k resistor and 100 µF capacitor.
The above values give a time constant
($R \times C$) of approximately 50 seconds.

(If the LM311 is used, remember it uses an
open collector output circuit – so connect a
pull-up resistor from the output to the positive supply.)

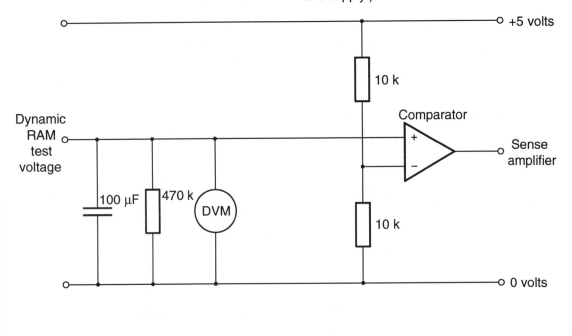

Mask Programmed ROM

The *mask programmed ROM* represents a compromise between the cost effectiveness of mass production and the need for customer-specific memory content. With this type of device, production proceeds normally until the final stage, at which point a custom designed metal *mask* is overlaid onto the IC.

Each bit in the memory array contains a MOS transistor. The mask is used to selectively connect particular transistors, while leaving others open-circuited, thus defining the memory content of the ROM. Figure 2.29 shows a simple memory array, with a mask used to define connections between particular rows and columns.

When a particular memory location (or row) is selected, all unconnected columns are pulled up to logic 1 by the FET load resistors, while connected columns are actively pulled low. For example if row N is selected, the three transistors connected to this row are enabled. The resulting output from the ROM is 0001_2.

This manufacturing process is only suitable for large volume production, due to the cost of manufacturing the mask. Where production runs of several thousand units are possible, and the ROM content is certain to remain unchanged, mask programmed ROM is the cheapest available form of non-volatile memory.

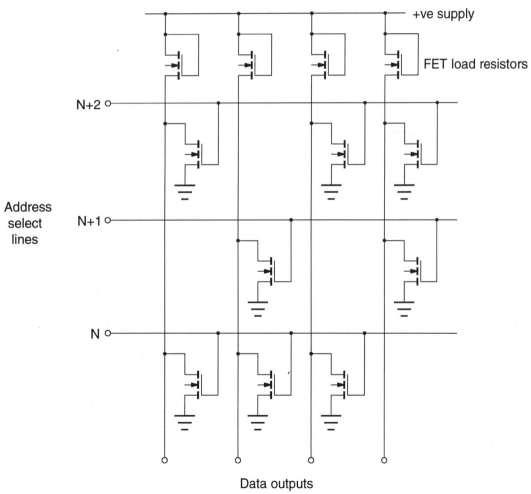

Fig. 2.29 Simple mask programmed ROM.

EPROM

EPROMs are one of the most widely used forms of non-volatile memory. This is due to their relatively low cost, wide availability and the ability to be programmed and erased using inexpensive equipment.

The operation of EPROMs is based on the controlled storage of electric charge at particular sites on the surface of a silicon wafer. These storage areas are surrounded by insulating material of extremely high resistance, which allows the charge to be retained for periods of twenty years or more!

Under certain conditions, such as exposure to ultraviolet light (high energy electromagnetic radiation), these stored charges are released, allowing the device to be re-used. This erasure process is typically performed in about twenty minutes. Figure 2.30 shows the structure of a typical EPROM cell, which is formed from an FET with a *floating gate* arrangement.

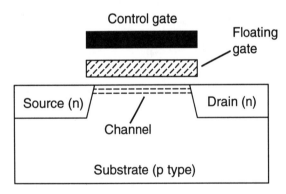

Fig. 2.30 Typical EPROM cell structure.

The control gate of the FET is connected to the address selection circuitry in the normal way. Transistor operation then depends on the presence or absence of stored charge on the floating gate area.

Under normal circumstances, conduction occurs in an n-channel FET when the gate is made positive, which attracts electrons into the area under the gate. In an enhancement type device this causes a temporary conducting channel to be formed between the source and drain.

If the floating gate contains a negative charge, due to the presence of trapped electrons, this tends to mask the effect of the positive potential on the gate terminal. A greater positive bias is then required on the control gate to switch the transistor on, as shown in figure 2.31.

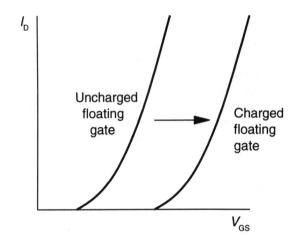

Fig. 2.31 Threshold shifting in a charged EPROM cell.

The internal circuitry of the EPROM is normally arranged so that an uncharged cell reads as a logic 1, while a charged cell appears as a logic 0. (So each location in a blank EPROM contains FF_{16}).

To inject electrons onto the floating gate, the control gate is connected to a positive programming voltage, which causes electrons to be attracted into the channel. If the applied electric field is sufficiently high, electrons cross the insulating barrier (due to an *avalanche* effect) onto the floating gate, where they become trapped. Each location is programmed and then verified in sequence, using a special programming unit.

Erasure of the EPROM is due to the *photoelectric effect*, in which electrons are temporarily excited by the absorption of high energy photons. These excited electrons have sufficient energy to cross the insulating barrier and escape. To erase the EPROM, ultraviolet light is allowed to shine onto the surface of the silicon, via a transparent quartz window (glass is opaque to UV light). A purpose-designed UV light exposure unit is used to erase the EPROM, prior to re-programming.

EEPROM

The storage unit of an EEPROM is similar in structure to that of an EPROM (as shown in figure 2.30), although the methods of programming and erasure are different. In both cases, charge storage on a floating gate is used to modify the switching characteristics of an FET.

In the EEPROM (also known as an *EAROM* or *electrically alterable ROM*), negative charge is stored in the floating gate region by a quantum mechanical effect called *electron tunnelling*. This process allows electrons to 'hop over' a thin insulating region between two conductors, as long as the electrons have sufficient energy.

One way to think of this is that there is a small *uncertainty* in the physical position of each electron at any time. This uncertainty increases with higher particle energies, to the point where there is a finite probability of the particle appearing on the other side of the insulating barrier! This very strange effect is based on Werner Heisenberg's *uncertainty principle* which states that the momentum and position of a particle cannot both be accurately determined at the same time.

As in the EPROM, the control gate is made positive to cause electrons to be attracted toward the floating gate region. Unlike EPROMs, modern EEPROM devices do not require an externally applied programming voltage. This allows them to be programmed in-situ, without any need for a special programming unit.

Another advantage of EEPROMs is that the tunnelling effect used to charge a storage cell may be reversed. By making the p-type substrate positive, while grounding the control gate, electrons on the floating gate are encouraged to tunnel back into the channel region. This allows storage cells to be reprogrammed without the need for erasure with ultraviolet light.

Repeated programming and erasure of EEPROM storage cells causes damage to the insulating layers in the gate region, due to the high electric fields applied. EEPROM design is a compromise between fast programming times (high electric fields) and long life. *FLASH* type EEPROMs overcome some of these problems by allowing the entire memory array to be erased simultaneously (in a flash), using lower field strengths.

Practical Memory Devices

This section examines a small selection of commercially available memory devices. Simplified timing diagrams, together with any special interfacing requirements are also considered.

The 6264 (8K × 8) Static RAM

The 6264 is available from a wide range of semiconductor manufacturers. Figure 2.32 shows the pin connections for this device.

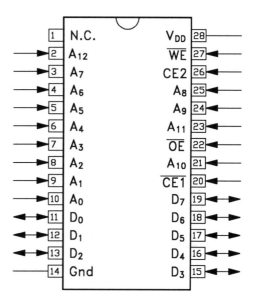

Fig. 2.32 6264 static RAM.

Examining the pin layout for this device gives the following information.

- 8 bit data width, allowing direct connection to the data bus of an 8 bit microprocessor.
- 13 address lines, allowing direct connection to A_{12}–A_0. (More significant address lines are connected to the address decoding logic).
- Dual chip enable signals, allowing active high and/or active low selection.
- Read or write selection using active low *output enable* and *write enable* signals.
- Single (5 volt) power supply.

Component catalogues normally offer a range of options on the basic type, including various access times and special low power (CMOS) devices, intended for battery-backed operation. In general, ICs with the fast access times or low power consumption are more expensive, so the ideal solution is to choose a device which is 'just' good enough. This decision is based on an examination of the timing of relevant microcomputer signals including,

- the availability of a valid address on the address bus,
- activation of the chip enable signal, caused by the address decoding logic,
- timing of relevant control signals, such as read, write, memory request, write enable and output enable,
- timing restraints imposed by the memory itself, such as the access time (the time taken for the memory to respond with valid data once all inputs are correct).

Figure 2.33 shows a simplified timing diagram for a memory read cycle.

In this example, the chip enable signals are assumed to become valid shortly after the address bus stabilises. The exact time delay may be determined by examining the propagation delay introduced by the address decoding logic.

The output enable signal is assumed to be derived from the appropriate control bus signals. (Example memory interface circuits were given in figures 1.15 and 1.16.)

The timing diagram for a memory write cycle is similar to that seen previously, except that write enable is active, rather than output enable. Figure 2.34 shows a typical example.

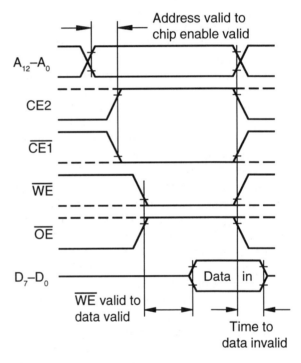

Fig. 2.34 Memory write timing.

The write cycle may be terminated either by the rising edge of the write enable signal, or by either of the chip enable signals becoming invalid. It is this transition that causes the information on the data bus to be latched by the memory.

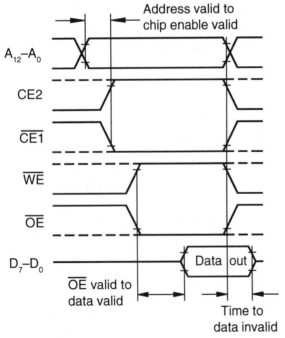

Fig. 2.33 Memory read timing.

Practical Exercise   Memory Read/Write Timing Diagrams

Introduction

The aim of this exercise is to study the timing of all relevant address data and control signals, during memory read and memory write cycles. This information may be displayed on a graph with the actual timing of all signals shown.

A logic analyser is the ideal tool for this task, although a dual channel oscilloscope may suffice. (The relative timing of up to three signals may be obtained by triggering the oscilloscope externally from the first signal, while observing the other two.)

Z80 Test Programs

```
8000H C3 00 80     JP 8000H
```

This program performs an op. code fetch from address 8000_{16}, followed by memory read operations from locations 8001_{16} and 8002_{16} in a repeating sequence. (The data bus contains $C3_{16}$, followed by 00_{16} and then 80_{16} at these times.)

```
8000H 32 00 40     LD (4000H),A
8003H C3 00 80     JP 8000H
```

The above program produces similar results to the first program but also generates a memory write cycle to address 4000_{16} (with the content of the accumulator present on the data bus.)

6502 Test Programs

```
8000H 4C 00 80     JMP $8000
```

This program performs an op. code fetch from address 8000_{16}, followed by memory read operations from locations 8001_{16} and 8002_{16} in a repeating sequence. (The data bus contains $4C_{16}$, followed by 00_{16} and then 80_{16} at these times.)

```
8000H 8D 00 40     STA $4000
8003H 4C 00 80     JMP $8000
```

The above program produces similar results to the first program but also generates a memory write cycle to address 4000_{16} (with the content of the accumulator present on the data bus.)

Results

Draw graphs to show the timing of all relevant signals during memory read and memory write cycles.

Signals of interest include,

- address bus and the memory chip enable signal,
- control bus and the write enable and output enable signals,
- data bus content, timing and direction.

TMS4C1024 Dynamic RAM

Dynamic RAMs, such as the TMS4C1024 (which is produced by Texas Instruments), are typically used where large memory capacities are required. This is particularly true in modern personal and business computing, where the increasing complexity of software has demanded progressively higher operating speeds and larger address spaces.

The TMS4C1024 is arranged internally as 1,048,576 locations, each of which is 1 bit wide. For this reason dynamic RAMs are normally purchased in pre-assembled modules, containing one RAM chip for each data bus line.

As mentioned earlier, the address inputs are normally *multiplexed*, which allows the upper and lower halves of the address to be applied to the same pins at different times. Row and column *address strobe* signals ($\overline{RAS}$ and $\overline{CAS}$) are then used to latch the address information within the RAM. Figure 2.35 shows a typical memory read timing diagram.

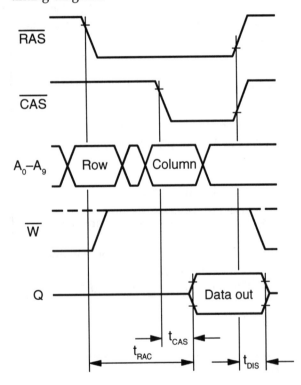

Fig. 2.35 Dynamic RAM memory read timing.

As can be seen, a read cycle is indicated by a high logic level on the $\overline{W}$ line (active low write signal), while 1 to 0 transitions on $\overline{RAS}$ and $\overline{CAS}$ are used to latch the two halves of the address bus. Shortly after the address has been latched, the data output (on pin Q) becomes valid. The data output is disabled following a 0 to 1 transition on $\overline{CAS}$.

The sequence of events during a write cycle is similar, except that data is applied externally to the D input and the $\overline{W}$ signal is active (at logic 0).

A memory refresh cycle may be carried out by performing a read operation to each of the 512 rows within the dynamic RAM. The maximum time interval between refresh cycles is stated by the manufacturer to be 8 ms, to ensure data retention. Although the TMS4C1024 supports three different refresh methods, the easiest to understand is the $\overline{RAS}$-only technique. This is illustrated in figure 2.36.

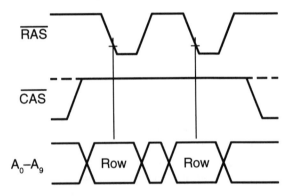

Fig. 2.36 Dynamic RAM $\overline{RAS}$-only refresh.

During $\overline{RAS}$-only refresh, the row address is applied and the falling edge of $\overline{RAS}$ causes that row to be refreshed. Output buffers are disabled due to the logic 1 applied to the $\overline{CAS}$ input.

The need for address bus multiplexing and regular memory refresh cycles complicates the interfacing between dynamic RAM and the microprocessor. For this reason a *dynamic RAM controller* IC is normally used to simplify the system design.

The following pages give introductory information on the TMS4C1024 dynamic RAM, an 8 bit dynamic RAM module, and a typical dynamic RAM controller IC. (Reproduced courtesy of Texas Instruments.)

TMS4C1024, TMS4C1025, TMS4C1027
1 048 576-BIT DYNAMIC RANDOM-ACCESS MEMORIES

SMGS024F — MAY 1986 — REVISED NOVEMBER 1990

This Data Sheet Is Applicable to All TMS4C1024/5/7s Symbolized with Revision "D" and Subsequent Revisions as Described on Page 5-62.

- 1 048 576 × 1 Organization

- Single 5-V Supply (10% Tolerance)

- Performance Ranges:

	ACCESS TIME $t_{a(R)}$ (t_{RAC}) (MAX)	ACCESS TIME $t_{a(C)}$ (t_{CAC}) (MAX)	ACCESS TIME $t_{a(CA)}$ (t_{CAA}) (MAX)	READ OR WRITE CYCLE (MIN)
TMS4C1024-60	60 ns	15 ns	30 ns	110 ns
TMS4C1024-70	70 ns	18 ns	35 ns	130 ns
TMS4C102_-80	80 ns	20 ns	40 ns	150 ns
TMS4C102_-10	100 ns	25 ns	45 ns	180 ns
TMS4C102_-12	120 ns	30 ns	55 ns	220 ns

- TMS4C1024 – Enhanced Page Mode Operation for Faster Memory Access
  - Higher Data Bandwidth than Conventional Page-Mode Parts
  - Random Single-Bit Access Within a Row With a Column Address

- TMS4C1025 – 4-Bit Nibble Mode Operation
  - Four Sequential Single-Bit Access Within a Row By Toggling $\overline{CAS}$

- TMS4C1027– Static Column Decode Mode Operation
  - Random Single-Bit Access Within a Row With Only a Column Address Change

- One of TI's CMOS Megabit DRAM Family, Including TMS44C256 – 256K × 4 Enhanced Page Mode

- $\overline{CAS}$-Before-$\overline{RAS}$ Refresh

- Long Refresh Period . . . 512-Cycle Refresh in 8 ms (Max)

- 3-State Unlatched Output

- Low Power Dissipation

- Texas Instruments EPIC™ CMOS Process

- All Inputs/Outputs and Clocks Are TTL Compatible

- Operating Free-Air Temperature Range . . . 0°C to 70°C

- High-Reliability Plastic 18-Pin 300-Mil-Wide DIP, 20/26 J-Lead Surface Mount (SOJ), 20/26 Thin J-Lead Surface Mount (ThinSOJ) or 20-Pin Zig-Zag In-line (ZIP) Packages

- Operations of TI's Megabit CMOS DRAMs Can Be Controlled by TI's SN74ALS6301 and SN74ALS6302 Dynamic RAM Controllers

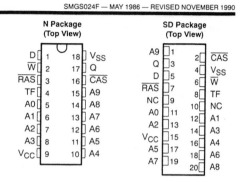

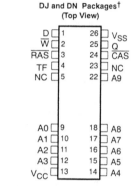

†The packages shown here are for pinout reference only. The DJ package is actually 75% of the length of the N package.

PIN NOMENCLATURE	
A0-A9	Address Inputs
$\overline{CAS}$	Column-Address Strobe
D	Data In
NC	No Connection
Q	Data Out
$\overline{RAS}$	Row-Address Strobe
TF	Test Function
$\overline{W}$	Write Enable
V_{CC}	5-V Supply
V_{SS}	Ground

EPIC is a trademark of Texas Instruments Incorporated.

POST OFFICE BOX 1443 • HOUSTON, TEXAS 77001

Fig. 2.37(a) TMS4C1024 – 1,048,576 × 1 bit dynamic RAM (courtesy of Texas Instruments).

TMS4C1024, TMS4C1025, TMS4C1027
1 048 576-BIT DYNAMIC RANDOM-ACCESS MEMORIES

SMGS024F — MAY 1986 — REVISED NOVEMBER 1990

description

The TMS4C1024, TMS4C1025, and TMS4C1027 are high-speed, 1 048 576-bit dynamic random access memories, organized as 1 048 576 words of one bit each. They employ state-of-the-art EPIC™ (Enhanced Process Implanted CMOS) technology for high performance, reliability, and low power at a low cost.

These devices feature maximum $\overline{RAS}$ access times of 60 ns, 70 ns, 80 ns, 100 ns, and 120 ns. Maximum power dissipation is as low as 305 mW operating and 11 mW standby on 120 ns devices.

The EPIC technology permits operation from a single 5-V supply, reducing system power supply and decoupling requirements, and easing board layout. I_{CC} peaks are 140 mA typical, and a -1-V input voltage undershoot can be tolerated, minimizing system noise considerations.

All inputs and outputs, including clocks, are compatible with Series 74 TTL. All addresses and data-in lines are latched on-chip to simplify system design. Data out is unlatched to allow greater system flexibility.

The TMS4C102_ are offered in an 18-pin plastic dual-in-line (N suffix) package, a 20/26 J-lead plastic surface mount SOJ (DJ suffix) package, a 20/26 J-lead thin plastic surface mount SOJ (DN suffix), and a 20-pin zig-zag in-line (SD suffix) package. The TMS4C1024-60 and TMS4C1024-70 are available in the 20/26 J-lead plastic surface mount SOJ (DJ suffix) only. These packages are characterized for operation from 0°C to 70°C.

operation

enhanced page mode (TMS4C1024)

Enhanced page-mode operation allows faster memory access by keeping the same row address while selecting random column addresses. The time for row-address setup and hold and address multiplex is thus eliminated. The maximum number of columns that may be accessed is determined by the maximum $\overline{RAS}$ low time and the $\overline{CAS}$ page cycle time used. With minimum $\overline{CAS}$ page cycle time, all 1024 columns specified by column addresses A0 through A9 can be accessed without intervening $\overline{RAS}$ cycles.

Unlike conventional page-mode DRAMs, the column-address buffers in this device are activated on the falling edge of $\overline{RAS}$. The buffers act as transparent or flow-through latches while $\overline{CAS}$ is high. The falling edge of $\overline{CAS}$ latches the column addresses. This feature allows the TMS4C1024 to operate at a higher data bandwidth than conventional page-mode parts, since data retrieval begins as soon as column address is valid rather than when $\overline{CAS}$ transitions low. This performance improvement is referred to as "enhanced page mode". Valid column address may be presented immediately after row address hold time has been satisfied, usually well in advance of the falling edge of $\overline{CAS}$. In this case, data is obtained after $t_{a(C)}$ max (access time from $\overline{CAS}$ low), if $t_{a(CA)}$ max (access time from column address) has been satisfied. In the event that column addresses for the next page cycle are valid at the time $\overline{CAS}$ goes high, access time for the next cycle is determined by the later occurrence of $t_{a(C)}$ or $t_{a(CP)}$ (access time from rising edge of $\overline{CAS}$).

Texas Instruments

POST OFFICE BOX 1443 • HOUSTON, TEXAS 77001

Fig. 2.37(b)   TMS4C1024 – 1,048,576 × 1 bit dynamic RAM (courtesy of Texas Instruments).

TMS4C1024, TMS4C1025, TMS4C1027
1 048 576-BIT DYNAMIC RANDOM-ACCESS MEMORIES

SMGS024F — MAY 1986 -- REVISED NOVEMBER 1990

nibble mode (TMS4C1025)

Nibble-mode operation allows high-speed read, write, or read-write-modify-write access of 1 to 4 bits of data. The first bit is accessed in the normal manner with read data coming out at $t_{a(C)}$ time as long as $t_{a(R)}$ and $t_{a(CA)}$ are satisfied. The next sequential bits can be read or written by cycling $\overline{CAS}$ while $\overline{RAS}$ remains low. The first bit is determined by the row and column addresses, which need to be supplied only for the first access. Row A9 and column A9 provide the two binary bits for initial selection, with row A9 being the least-significant address and column A9 being the most significant. Thereafter, the falling edge of $\overline{CAS}$ will access the next bit of the circular 4-bit nibble in the following sequence.

Data written in a sequence of more than 4 consecutive cycles shall be capable of being read back without exiting from the nibble mode. In a sequence of consecutive nibble-mode cycles the control of the high-impedance state for the data out (Q) pin is determined by each individual cycle. This facilitates fully mixed nibble-mode cycles (e.g., read/write/read-modify-write/read etc.).

static column decode mode (TMS4C1027)

The static column decode mode of operation allows high-speed read, write, or read-modify-write by reducing the number of required signal setup, hold, and transition timings. This is achieved by first addressing the row and column in the normal manner, but after the first access, maintaining $\overline{CAS}$ low. Subsequently changing the column address produces valid data at $t_{a(CA)}$. The first bit is accessed in the normal manner with read coming out at $t_{a(R)}$ time. Similarly, write or read-modify-write cycle times can be achieved with appropriate toggling of $\overline{W}$. The addresses are latched during the write operation, and remain latched unitl $\overline{CAS}$ or $\overline{W}$ no longer remains low.

address (A0 through A9) (TMS4C1024, TMS4C1025)

Twenty address bits are required to decode 1 of 1 048 576 storage cell locations. Ten row-address bits are set up on inputs A0 through A9 and latched onto the chip by the row-address strobe ($\overline{RAS}$). The ten column-address bits are set up on pins A0 through A9 and latched onto the chip by the column-address strobe ($\overline{CAS}$). All addresses must be stable on or before the falling edges of $\overline{RAS}$ and $\overline{CAS}$. $\overline{RAS}$ is similar to a chip enable in that it activates the sense amplifiers as well as the row decoder. $\overline{CAS}$ is used as a chip select activating the output buffer, as well as latching the address bits onto the column-address buffer.

address (A0 through A9) (TMS4C1027)

Twenty address bits are required to decode 1 of 1 048 576 storage cell locations. Ten row-address bits are set up on pins A0 through A9 and latched onto the chip by the row-address strobe ($\overline{RAS}$). The ten column-address bits are set up on pins A0 through A9. Row addresses must be stable on or before the falling edges of $\overline{RAS}$. $\overline{RAS}$ is similar to a chip enable in that it activates the sense amplifiers as well as the row decoder. In a write cycle, the later of $\overline{CAS}$ or $\overline{W}$ latches the column address bits.

write enable ($\overline{W}$)

The read or write mode is selected through the write enable ($\overline{W}$) input. A logic high on the $\overline{W}$ input selects the read mode and a logic low selects the write mode. The write-enable terminal can be driven from the standard TTL circuits without a pullup resistor. The data input is disabled when the read mode is selected. When $\overline{W}$ goes low prior to $\overline{CAS}$ (early write), data out will remain in the high-impedance state for the entire cycle, permitting common I/O operation.

TEXAS
INSTRUMENTS

POST OFFICE BOX 1443 • HOUSTON, TEXAS 77001

5-25

Fig. 2.37(c) TMS4C1024 – 1,048,576 × 1 bit dynamic RAM (courtesy of Texas Instruments).

TMS4C1024, TMS4C1025, TMS4C1027
1 048 576-BIT DYNAMIC RANDOM-ACCESS MEMORIES

SMGS024F — MAY 1986 — REVISED NOVEMBER 1990

data in (D)

Data is written during a write or read-modify-write cycle. Depending on the mode of operation, the falling edge of $\overline{CAS}$ or $\overline{W}$ strobes data into the on-chip data latch. In an early write cycle, $\overline{W}$ is brought low prior to $\overline{CAS}$ and the data is strobed in by $\overline{CAS}$ with setup and hold times referenced to this signal. In a delayed-write or read-modify-write cycle, $\overline{CAS}$ will already be low, thus the data will be strobed in by $\overline{W}$ with setup and hold times referenced to this signal.

data out (Q)

The three-state output buffer provides direct TTL compatibility (no pullup resistor required) with a fanout of two Series 74 TTL loads. Data out is the same polarity as data in. The output is in the high-impedance (floating) state until $\overline{CAS}$ is brought low. In a read cycle the output becomes valid after the access time interval $t_{a(C)}$ that begins with the negative transition of $\overline{CAS}$ as long as $t_{a(R)}$ and $t_{a(CA)}$ are satisfied. The output becomes valid after the access time has elapsed and remains valid while $\overline{CAS}$ is low; $\overline{CAS}$ going high returns it to a high-impedance state. In a delayed-write or read-modify-write cycle, the output will follow the sequence for the read cycle.

refresh

A refresh operation must be performed at least once every eight milliseconds to retain data. This can be achieved by strobing each of the 512 rows (A0-A8). A normal read or write cycle will refresh all bits in each row that is selected. A $\overline{RAS}$-only operation can be used by holding $\overline{CAS}$ at the high (inactive) level, thus conserving power as the output buffer remains in the high-impedance state. Externally generated addresses must be used for a $\overline{RAS}$-only refresh. Hidden refresh may be performed while maintaining valid data at the output pin. This is accomplished by holding $\overline{CAS}$ at V_{IL} after a read operation and cycling $\overline{RAS}$ after a specified precharge period, similar to a $\overline{RAS}$-only refresh cycle.

$\overline{CAS}$-before-$\overline{RAS}$ refresh

$\overline{CAS}$-before-$\overline{RAS}$ refresh is utilized by bringing $\overline{CAS}$ low earlier than $\overline{RAS}$ [see parameter $t_{d(CLRL)R}$] and holding it low after $\overline{RAS}$ falls [see parameter $t_{d(RLCH)R}$]. For successive $\overline{CAS}$-before-$\overline{RAS}$ refresh cycles, $\overline{CAS}$ can remain low while cycling $\overline{RAS}$. The external address is ignored and the refresh address is generated internally. The external address is also ignored during the hidden refresh cycles.

power-up

To achieve proper device operation, an initial pause of 200 µs followed by a minimum of eight initialization cycles is required after full V_{CC} level is achieved.

test function pin

During normal device operation the TF pin must either be disconnected or biased at a voltage less than or equal to V_{CC}.

TEXAS
INSTRUMENTS

POST OFFICE BOX 1443 • HOUSTON, TEXAS 77001

Fig. 2.37(d)   TMS4C1024 – 1,048,576 × 1 bit dynamic RAM (courtesy of Texas Instruments).

TM024GAD8
1 048 576 BY 8-BIT
DYNAMIC RAM MODULE
SMMS108A — MARCH 1990 — REVISED NOVEMBER 1990

This Data Sheet is Applicable to All
TM024GAD8s Manufactured With
TMS4C1024s Symbolized With Revision "D"
and Subsequent Revisions.

- **TM024GAD8 . . . 1 048 576 × 8 Organization**

- **Single 5-V Supply (10% Tolerance)**

- **30-Pin Single-In-Line Package (SIP)**
  — Leadless Module for Use With Sockets

- **Utilizes Eight 1-Megabit Dynamic RAMs in**
  Plastic Small-Outline J-Lead (SOJ)
  Packages

- **Long Refresh Period . . . 8 ms**
  (512 Cycles)

- **All Inputs, Outputs, Clocks Fully TTL**
  Compatible

- **3-State Output**

- **Performance of Unmounted RAMs:**

	ACCESS TIME t_{RAC} (MAX)	ACCESS TIME t_{CAC} (MAX)	READ OR WRITE CYCLE (MIN)	V_{CC} TOLERANCE
TMS4C1024-6	60 ns	15 ns	110 ns	5%
TMS4C1024-70	70 ns	18 ns	130 ns	10%
TMS4C1024-80	80 ns	20 ns	150 ns	10%
TMS4C1024-10	100 ns	25 ns	180 ns	10%

- **Common $\overline{CAS}$ Control for Eight Common**
  Data-In and Data-Out Lines

- **Low Power Dissipation**

- **Operating Free Air Temperature**
  . . . 0°C to 70°C

description

The TM024GAD8 is a 8192K (dynamic) random-access memory module organized as 1 048 576 × 8 in a 30-pin single-in-line (SIP) module. The TM024GAD8 is composed of eight TMS4C1024DJ, 1 048 576 × 1-bit dynamic RAMs, each in 20/26-lead plastic small-outline J-lead packages (SOJ), mounted on a substrate together with decoupling capacitors.

AD Single-In-Line Package
(Top View)

Pin	Signal
1	V_{CC}
2	$\overline{CAS}$
3	DQ1
4	A0
5	A1
6	DQ2
7	A2
8	A3
9	V_{SS}
10	DQ3
11	A4
12	A5
13	DQ4
14	A6
15	A7
16	DQ5
17	A8
18	A9
19	NC
20	DQ6
21	$\overline{W}$
22	V_{SS}
23	DQ7
24	NC
25	DQ8
26	NC
27	$\overline{RAS}$
28	NC
29	NC
30	V_{CC}

PIN NOMENCLATURE	
A0-A9	Address Inputs
$\overline{CAS}$	Column-Address Strobe
DQ1-DQ8	Data In/Data Out
NC	No Connection
$\overline{RAS}$	Row-Address Strobe
V_{CC}	5-V Supply
V_{SS}	Ground
$\overline{W}$	Write Enable

TEXAS
INSTRUMENTS

POST OFFICE BOX 1443 • HOUSTON, TEXAS 77001

6-7

Fig. 2.38(a) TM024GAD8 – 1,048,576 × 8 bit dynamic RAM module (courtesy of Texas Instruments).

TM024GAD8
1 048 576 BY 8-BIT
DYNAMIC RAM MODULE
SMMS108A — MARCH 1990 — REVISED NOVEMBER 1990

functional block diagram

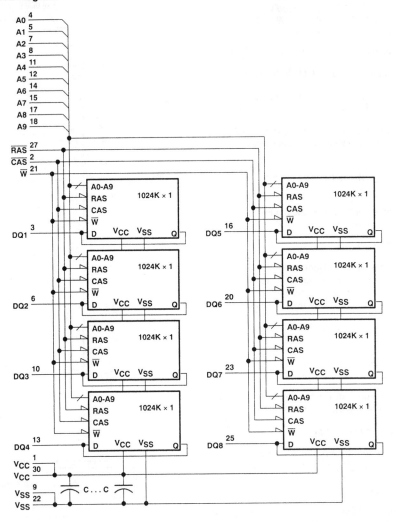

POST OFFICE BOX 1443   •   HOUSTON, TEXAS 77001

Fig. 2.38(b)   TM024GAD8 – 1,048,576 × 8 bit dynamic RAM module (courtesy of Texas Instruments).

SN74ACT4503
DYNAMIC RAM CONTROLLER

D3132, SEPTEMBER 1988 – REVISED MAY 1989

- Inputs are TTL- and CMOS-Voltage Compatible

- Controls Operation of 64K, 256K, and 1M Dynamic RAMs

- Creates Static RAM Appearance

- One Package Contains Address Multiplexer, Refresh Control, and Timing Control

- Directly Addresses and Drives Up to 4 Banks of Memory

- Operates from Microprocessor Clock
  - No Crystals, Delay Lines, or RC Networks
  - Eliminates Arbitration Delays

- Refresh May Be Internally or Externally Initiated

- Versatile
  - Strap-Selected Refresh Rate
  - Synchronous, Predictable Refresh
  - Selection of Distributed, Transparent, and Cycle-Steal Refresh Modes
  - Interfaces Easily to Popular Microprocessors
  - Asynchronous $\overline{\text{RESET}}$
  - Choice of CLK Polarity on Refresh/Access Arbitration

- High-Performance Si-Gate CMOS Technology

- Strap-Selected Refresh Frequencies for Microprocessor/Memory Speed Matching

- Ability to Synchronize or Interleave Controller with the Microprocessor System (Including Multiple Controllers)

- 3-State Outputs Allow Multiport Memory Configuration

- Performance Range: 100 ns ALE low to $\overline{\text{CAS}}$ low

- Functionally Compatible with TMS4500A/B and with THCT4502B

- Available in Plastic and Ceramic Chip Carriers in Addition to Plastic and Ceramic DIPs

- Dependable Texas Instruments Quality and Reliability

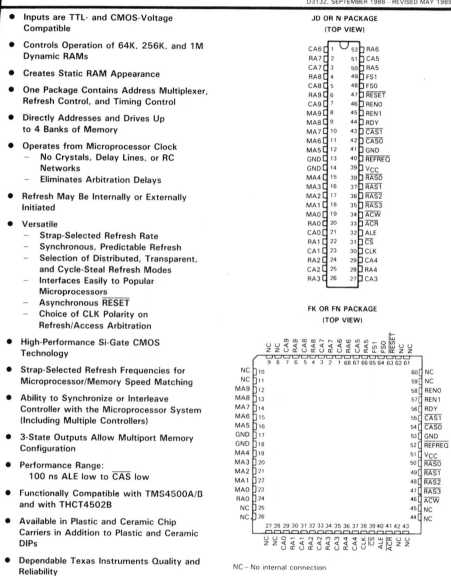

NC – No internal connection

POST OFFICE BOX 655303 • DALLAS, TEXAS 75265

10-35

Fig. 2.39(a) SN74ACT4503 – Dynamic RAM controller (courtesy of Texas Instruments).

SN74ACT4503
DYNAMIC RAM CONTROLLER

description

The 'ACT4503 is a monolithic DRAM system controller providing address multiplexing, timing, control, and refresh/access arbitration functions to simplify the interface of dynamic RAMs to microprocessor systems.

The controller contains an 20-bit multiplexer that generates the address lines for the memory device from the 20 system address bits and provides the strobe signals required by the memory to decode the address. A 10-bit refresh counter generates up to 1024 row addresses required to refresh.

A refresh timer is provided to generate the necessary timing to refresh the dynamic memories and ensure data retention.

for complete data sheet

The complete version of this data sheet and application information can be found in the *Cache Memory Management Data Book*, Literature #SCAD002. To obtain a copy of this data book, contact your local TI sales representative or call the TI Customer Response Center at 1-800-223-3200.

logic symbol†

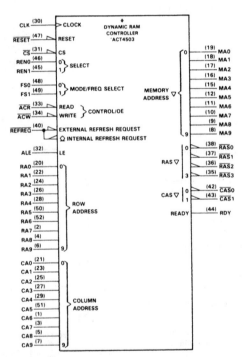

†This symbol is in accordance with ANSI/IEEE Std. 91-1984.
Pin numbers shown are for JD and N packages.

TEXAS
INSTRUMENTS
POST OFFICE BOX 655303 • DALLAS, TEXAS 75265

Fig. 2.39(b)   SN74ACT4503 – Dynamic RAM controller (courtesy of Texas Instruments).

Cache Memory Systems

Specifications of microcomputers with large memory capacities frequently include *cache memory*, which may be either internal or external to the microprocessor. Although this is an advanced topic, it is useful to understand the function of a cache system, and the reasons for its use.

As mentioned previously, dynamic RAM is generally available in higher memory capacities than static RAM. Thus to reduce system complexity, large capacity memory systems are based almost exclusively on dynamic RAM.

Another trend in modern computing is that microprocessors are becoming available with higher clock frequencies. (At the time of writing, clock frequencies are approaching 100 megahertz.) Advances in dynamic RAM access times have been slow in comparison, with the result that access times are one of the major factors influencing system performance.

A cache system uses a small area of very fast static RAM to hold frequently used data. Whenever a memory read operation takes place, the microprocessor first checks the cache to see if it contains the required data. (This is called a *cache hit*.) In the event of a hit, the required data is provided by the cache memory, which is much faster than reading from main memory. If the data is not found (which is called a *cache miss*), then the memory read operation proceeds as normal from dynamic RAM. The content of the cache is then updated so that any further memory read operation from this address, will be supplied by the cache. Internal mechanisms in the cache ensure that up to date information is retained, while old data is discarded.

The content of each memory location in the cache consists of a *tag*, plus the actual data for that location. To understand the function of the tag, consider an example of a computer with one megabyte of dynamic RAM and a 4K cache, which uses an 8 bit tag. During a memory read, the lower 12 bits of the address bus are applied directly to the cache, selecting one of the 4K cache locations. The upper 8 bits of the address bus are then compared with the tag. If these match, then a cache hit is said to have occurred and the cache supplies the requested data at high speed.

2764 (8K × 8) EPROM

The 2764 EPROM is a commonly available device, which has a similar pin configuration to the 6264 static RAM seen earlier. Figure 2.40 shows the pin layout for this device.

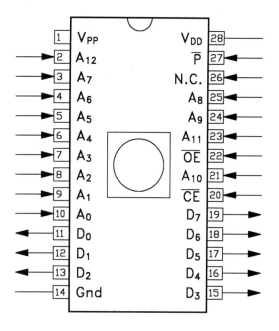

Fig. 2.40 2764 (8K × 8) EPROM.

Read mode operation is, as expected, similar to the 6264. The only differences are the use of a single chip enable input and the absence of a write enable control signal (for obvious reasons).

Programming of the 2764 is normally performed using an EPROM programmer, which requires removal of the EPROM from the computer. The programming cycle is actually quite complex, requiring special programming voltages, in addition to the normal 5 volt supply. With the 2764, the positive supply pin (V_{DD}) must be raised to 6 volts, while the programming voltage pin (V_{PP}) is connected to 12.5 volts.

Once V_{DD} and V_{PP} are connected to the required voltage levels, each location may then be programmed and verified. The programming cycle for a single location involves applying the required address and data inputs, after which the program ($\overline{P}$) pin is held low for a fixed period of time (1 ms in the case of the 2764).

Following each program pulse, the content of the most recently programmed location is compared with the required data. If the location has programmed successfully then the programmer moves on to program the next location. In the event of a verification failure then another program pulse is applied. After a certain number of retries, if the location has still not programmed successfully, then the programmer software reports that the EPROM has failed to program.

Assuming that each location has been successfully programmed and verified, V_{PP} and V_{DD} are returned to 5 volts, after which the content of each location in the EPROM is again checked against the intended data. Figure 2.41 shows a typical program cycle, for a single location.

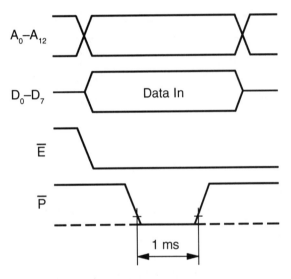

Fig. 2.41 2764 EPROM program cycle timing.

The time taken to program an entire EPROM may easily be estimated, based on a knowledge of the typical programming time for a single location and the total number of memory locations. With the 2764, assuming 1 ms per location (and 8096 locations), the programming time is about 8.1 seconds.

(Note that older EPROMs, such as the 2716 ($2K \times 8$) used fixed length programming pulses of 50 mS, and used a single programming voltage of approximately 21 volts. If any location failed to verify then no attempt was made to reprogram that location.)

As EPROM sizes have increased, program pulse widths have been steadily reduced in order to keep device programming times within acceptable limits. More recent EPROMs, such as the 27512 ($64K \times 8$) use program pulses of only 0.1 ms duration. Assuming that each location programs at the first attempt, this results in a device program time of around 6.5 seconds.

Consider the design of the EPROM programmer itself (which is itself a special purpose microcomputer): The device to be programmed is normally inserted into a *zero insertion force* (ZIF) socket.

The address and data signals to the EPROM are usually provided by several input/output ports. With the 2764 considered here, 21 port bits are required for the address and data buses. A further two lines are needed for the chip enable and program signals, with a final output bit used to switch the 6 and 12.5 volt supplies on and off. These 24 port bits could be supplied using three 8 bit ports.

The voltage switching circuitry may be based on a transistor circuit, or may use a double pole relay, as shown in figure 2.42.

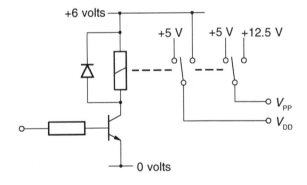

Fig. 2.42 EPROM power switching circuit.

EPROM erasure is performed using an ultraviolet lamp unit, which normally takes 15 to 20 minutes. In cases where the EPROM will never need to be erased, cheaper *one time programmable* (*OTP*) devices may be purchased. These are supplied in a plastic case without the transparent window, which reduces production costs.

Practical Exercise   Programming and Erasing EPROMs

Introduction

This exercise is intended to provide familiari-sation with programming and erasing of EPROMs. Timing of all relevant signals are also considered.

Method

a) Use an EPROM programmer to place a simple test program into the EPROM.

b) Insert the EPROM into an available socket and test the operation of the program.

c) Use an EPROM eraser to erase the EPROM, (and perform a blank check).

Equipment required

Suitable microcomputer, with an available ROM socket.
EPROM programmer.
EPROM eraser.
EPROM (to suit the above socket).
Storage oscilloscope or logic analyser.

Results and Conclusions

Sketch timing diagrams of all relevant signals during a typical program cycle. Compare your results with those expected.

How long does it take to correct an error in an EPROM based program, and why?

Practical Exercise   EEPROM Programming

Introduction

This exercise provides familiarisation with programming and erasing of EEPROMs. Their ease of use is compared with EPROMs.

Method

a) Insert an EEPROM into an available socket.

b) Use a suitable software routine, transfer test data to the EEPROM (using a data polling method).

c) Erase the EEPROM by copying new data.

Equipment required

Suitable microcomputer, with an available ROM socket.
EEPROM (to suit the above socket).
Storage oscilloscope or logic analyser.

Results and Conclusions

Sketch timing diagrams of all relevant signals during a typical program cycle. Compare your results with those expected.

Comment on the convenience of using EEPROMs, rather than EPROMs, where future modifications to the ROM content are likely.

2864 (8K × 8) EEPROM

The 2864 EEPROM is considered here to enable comparison with the 2764 EPROM, to which it is a direct alternative. Figure 2.43 shows the pin configuration of this device.

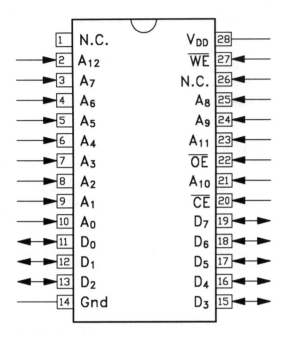

Fig. 2.43   2864 (8K × 8) EEPROM.

Memory read and write timing diagrams are virtually identical to those for the 6264 static RAM seen earlier. The most apparent difference is the use of a single (active low) chip enable in the 2864. Notice also that the 2864 lacks the programming voltage (V_{PP}) connection used by the 2764.

Following a memory write cycle, the EEPROM initiates an *internal write cycle* which is normally completed within 2 ms. During this time, if the content of the last programmed location is read, it is found to be different from the programmed data. Once programming is complete, the content of this location is found to be correct.

A simple algorithm known as *data polling* is often used when programming EEPROMs. Using this method, data is written to an EEPROM location. The program then executes a closed loop, waiting for the data content to become correct. This process is repeated for each location.

Example

Explain how data polling may be used to control the EEPROM programming sequence.

The EEPROM appears to be similar to a RAM integrated circuit. Memory read and memory write cycles are both possible. The main difference is that having written data to an EEPROM location, an internal write operation commences. During this time, the content of the programmed location differs from that intended.

The precise details may vary between devices but in many cases the most significant bit appears inverted. For example if 00_{16} is written into a location and immediately read back, the content may appear as 80_{16}. On completion of the internal write cycle a memory read cycle returns the expected result of 00_{16}.

A simple programming method may be developed, based on a data polling principle, as shown below.

Z80 solution...

```
          LD DE,SOURCE   ;Source (RAM)
          LD HL,DEST     ;Dest. (EEPROM)
          LD B,BYTES     ;Byte count
LP:       LD A,(DE)      ;Transfer data
          LD (HL),A      ;(DE) to (HL)
WT:       CP (HL)        ;Wait until
          JP NZ,WT       ;programmed
          INC HL
          INC DE
          DJNZ LP        ;Loop until all
          HALT           ;bytes written
```

6502 solution...

```
          LDX #BYTES     ;Byte count
LP:       LDA SOURCE,X   ;Source (RAM)
          STA DEST,X     ;Dest. (EEPROM)
WT:       CMP DEST,X     ;Wait until
          BNE WT         ;programmed
          DEX
          BNE LP         ;Loop until all
          BRK            ;bytes written
```

It should also be noted that some EEPROMs also have a *page write* facility, in which a group of bytes may be simultaneously programmed. This can be result in a significant saving in programming time, when compared to the single byte programming method just considered.

In EEPROMs with the page write facility, the internal write cycle does not begin immediately after a byte is wriiten. Instead, a short delay (typically 100 µs) occurs, during which further bytes may be written into an internal buffer. Each byte must be in the same 'page', meaning that the more significant address lines must remain constant during page write operations. A typical page size may be 64 bytes, although this varies between devices and manufacturers.

Summary

The main points covered in this chapter are:

- *Transistor transistor logic* (*TTL*) is a widely used logic family. A number of variations on the basic circuit are available including *Schottky* and *open collector* versions.
- *Metal oxide semiconductor field effect transistors* (*MOSFETs*) are used as the basis of several logic families. *P* and *n-channel* devices are used in *PMOS* and *NMOS* logic circuits. A combination of p and n channel transistors is used in *CMOS* ICs, resulting in lower power consumption and a wide operating voltage range.
- High speed computer circuitry is often based on *emitter coupled logic* (*ECL*).
- Computer memory is available in several forms, including *static RAM, dynamic RAM, mask programmed ROM, erasable programmable ROM* (*EPROM*) and *electrically erasable and programmable ROM* (*EEPROM*).
- Static RAM operation is based internally on the bistable, while dynamic RAM uses the storage of electric charge. Static RAM is simpler to use, but dynamic RAM is available in larger memory capacities, due to its simpler internal construction. Memory refresh of dynamic RAMs is normally performed by a dedicated controller IC.
- Mask programmed ROM is ideally suited to mass production, and in such situations it is the cheapest form of ROM. The content of the ROM is determined at the final stage of production, when a metal mask is overlaid onto the substrate.
- EPROMs are ideal for small volume production situations. A special purpose programming unit is used to program the device, while exposure to ultraviolet light is used to erase the memory content.
- A more flexible form of non-volatile memory is provided by EEPROMs. These may be programmed and erased without being removed from their sockets and do not require any special programming voltages.

Problems

1   State the bias condition required to cause conduction in the following types of transistor.

a) NPN
b) PNP
c) N-channel enhancement MOSFET
d) N-channel depletion MOSFET
e) P-channel enhancement MOSFET
f) P-channel depletion MOSFET

2   Describe, with the aid of suitable diagrams, the internal construction and operation of the following logic families.

a) TTL
b) NMOS
c) CMOS
d) ECL

3   Name and briefly describe, three different types of non-volatile semiconductor-based memory.

4   Compare the relative advantages and disadvantages of static and dynamic RAM.

5   Suggest (giving your reasons), a possible area of application for the following forms of non-volatile memory.

a) CMOS static RAM with battery backup
b) Mask programmed ROM
c) EPROM
d) OTP (one time programmable) EPROM
e) EEPROM

6   The floating gate of a fully charged EPROM cell contains 100,000 electrons. After 10 years it is found that 50% of the stored charge has leaked away.

a) On average, how many electrons escape each day?
b) If the charge on an electron is 1.6×10^{-19} coulombs, what flow of electric current does this represent?

7   Calculate the time taken to program a 2732 $(4K \times 8)$ EPROM, if the time taken to program a single location is,

a) 1 ms
b) 50 ms

3 The Stack and Interrupts

Aims

When you have completed this chapter you should be able to:

1 Understand how the stack, together with the stack pointer register may be used to store and later retrieve data.
2 Appreciate that subroutine operation is possible due to the last-in, first-out nature of the stack.
3 Use the stack to save and later restore the content of data registers used by the main program.
4 Understand the available types of interrupt signal and write interrupt service routines to make use of them.
5 Compare the relative merits of interrupts and polling.
6 Appreciate the need to prioritise interrupts.

The Stack and Stack Pointer

A subroutine may be thought of as a mini-program which performs some distinct function and which may be called from any part of the main program. When the subroutine is finished, the main program resumes at the point immediately after the *subroutine call* instruction.

Subroutines allow commonly used functions to be accessed from any part of the program, simply by calling the subroutine. An area of memory called the *stack* is used to hold the subroutine's return address. The use of subroutines results in shorter programs, which are easier to read and therefore less likely to contain errors. Figure 3.1 illustrates the sequence of events when a subroutine is called from the main program.

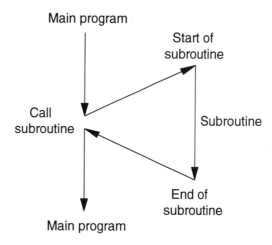

Fig. 3.1 Subroutine operation.

An area of RAM is allocated for the storage of temporary data, called the *stack*. The size of this area may depend on the quantity of available memory (as in the Z80), or it may be limited by the design of the microprocessor (with the 6502).

At any time the stack is likely to contain some data and will therefore be partially filled. The boundary between the used and unused areas is known as the *top of the stack*. A special memory pointer register known as the *stack pointer* indicates the position of this boundary. As information is stored or retrieved from the stack, the stack pointer alters so that it always points to the top of the stack.

In some microprocessors, such as the 6502, the stack pointer points to the first free or unused location in the stack. In others, such as the Z80, the stack pointer indicates the last used location. There are therefore slight differences in the detailed operation of the stack between different microprocessors.

Figure 3.2 shows the partially filled stack of a typical microcomputer, together with the associated stack pointer register.

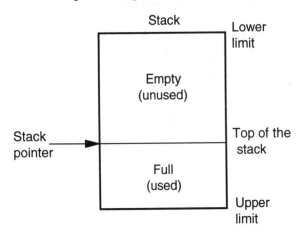

Fig. 3.2   The stack and the stack pointer.

When a jump to a subroutine is made, the address of the next instruction in the main program is stored on the stack. This allows a return to the main program to be made later by removing the return address from the stack and placing it in the program counter. Before considering the detailed operation of the stack on the Z80 and 6502 microprocessors, recall the instructions used to call or return from a subroutine, as shown in tables 3.1 and 3.2.

Instruction	Function
CALL address	Jump to subroutine.
CALL cc,address	Jump to subroutine on condition true.
RET	Return from subroutine.
RET cc	Return from subroutine on condition true.

Table 3.1   Z80 subroutine instructions.

Instruction	Function
JSR address	Jump to subroutine.
RTS	Return from subroutine.

Table 3.2   6502 subroutine instructions.

The Z80 Stack

The Z80 has a 16 bit stack pointer register which points to the most recently saved information on the stack. Figure 3.3 shows the sequence of events when a single byte is stored onto the Z80 stack.

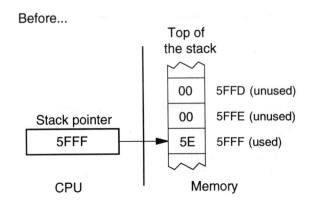

Fig. 3.3   Operation of the Z80 stack during data storage.

The stack pointer is first decremented, so that it points to an unused location, after which the required data is saved on the stack. This method prevents previously saved information from being overwritten when new data is saved. The stack 'grows' as more and more information is placed on the stack.

The procedure for removing data is the exact inverse of the above. The data is first removed from the location pointed to by the stack pointer, after which the stack pointer increments. The stack 'shrinks' as data is removed from the stack.

The 6502 Stack

The 6502 has an 8 bit stack pointer register, which points to the first unused location on the stack. The stack is limited in size to 256 bytes and is located in the memory range 0100_{16}–$01FF_{16}$ (*page one* of memory). Figure 3.4 illustrates the storage of data on the 6502 stack.

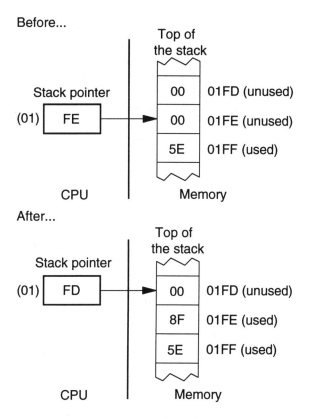

Before...

After...

Fig. 3.4 Operation of the 6502 stack during data storage.

The required data is saved on to the stack, after which the stack pointer is decremented. This method prevents previously saved information from being overwritten when new data is saved. The stack 'grows' as more and more information is placed on the stack.

The procedure for removing data is the exact inverse of the above. The stack pointer is incremented, after which the data byte is retrieved from the stack. The stack 'shrinks' as data is removed from the stack.

Subroutines and Nested Subroutines

When a *jump to a subroutine* is made, the return address to the main program is stored on the stack. This is the address of the instruction immediately following the subroutine call.

The return address is a 16 bit number, which is stored in two memory locations on the stack. In both the Z80 and 6502 microprocessors the most significant byte is stored first, followed by the least significant byte.

When the *return from subroutine* instruction is encountered this causes the two bytes on the top of the stack to be removed and placed into the program counter. The least significant byte of the program counter is removed first, followed by the most significant byte. Figure 3.5 shows the stack content when a call to a subroutine at address 5000_{16} is made from address 4000_{16}.

Before...

During...

After...

Fig. 3.5 The stack before, during and after a subroutine call.

As more information is stored on the stack, the content of the stack pointer register diminishes and the stack grows 'downwards' in memory.

When information is removed from the stack, it is always the most recently saved data which is on the top of the stack. For this reason the stack is often called a *last-in first-out* store.

This behaviour of the stack makes it possible for one subroutine to make a call to another subroutine. This is known as *subroutine nesting* and is illustrated by figure 3.6.

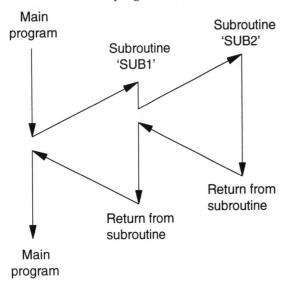

Fig. 3.6 Nested subroutines.

In this example the main program executes normally until a subroutine call to SUB1 is encountered. The return address to the main program is then stored on the stack and the first subroutine begins to run.

At some point during SUB1, a subroutine call to SUB2 occurs, which causes the return address to the first subroutine to be stored on the stack. The stack now contains two 16 bit return addresses. (The return address from the second subroutine is the most recently saved and is therefore on the top of the stack.)

When the second subroutine finishes, the return from subroutine instruction causes the top two bytes on the stack to be placed into the program counter. The first subroutine then resumes at the instruction following the second subroutine call. The stack once again contains a single 16 bit return address.

On completion of the first subroutine, the remaining two bytes are removed from the stack and placed into the program counter. At this point the main program resumes immediately after the first subroutine call. Figure 3.7 shows the content of the stack at each point of the nested subroutine program.

Before considering the use of the stack for temporary data storage, which is covered in the next section, it is useful to summarise the rules that allow reliable subroutines to be developed.

- Always call subroutines using the correct instruction (JSR or CALL), and place a return from subroutine at the end.
- Remember to set the stack pointer explicitly to an available area of RAM at the start of your program, unless other software (such as the monitor) does this automatically.
- Do not exceed the maximum storage capacity of the computer's stack. This is 256 bytes on the 6502, and is defined by RAM availability on the Z80.

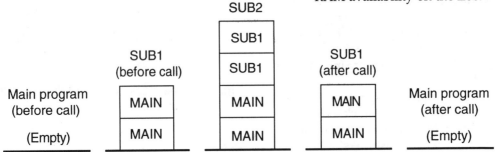

Fig. 3.7 The stack during nested subroutines.

Push and Pop Instructions

The stack also allows temporary data, such as the content of a register, to be saved and later restored. This is useful because once the content of a register has been temporarily stored on the stack, it may be used for some other purpose. On completion, the register may be restored to its previous value.

The Z80 and 6502 microprocessors provide a number of instructions which allow data to be placed or *pushed* onto the stack, and later *pulled* or *popped* from the stack. Tables 3.3 and 3.4 show the available push and pop instructions for the Z80 and 6502 microprocessors respectively

Push Instructions	Pop Instructions
PUSH AF	POP AF
PUSH BC	POP BC
PUSH DE	POP DE
PUSH HL	POP HL
PUSH IX	POP IX
PUSH IY	POP IY

Table 3.3   Z80 push and pop instructions.

Push Instructions	Pull Instructions
PHA	PLA
PHP	PLP

Table 3.4   6502 push and pull instructions.

The ability to save and restore microprocessor registers using the stack is useful in subroutines, and, as will be seen later, it is vital when dealing with interrupt requests!

Using push and pop instructions it is possible to write subroutines which do not alter any of the registers used by the main program. This is useful in the development of *subroutine libraries*, where member functions may be safely called without any knowledge of the subroutine's internal operation.

In this case, all of the data registers used by the subroutine are pushed onto the stack at the start of the subroutine, allowing them to recovered later. The subroutine then performs its normal function, after which the registers are restored to their previous state. Note that due to the last-in, first-out nature of the stack, registers must be popped in the reverse order to which they were originally pushed. Figure 3.8 shows the use of push and pop instructions to save and later restore all data registers used by a Z80 subroutine, while figure 3.9 gives the same information for the 6502.

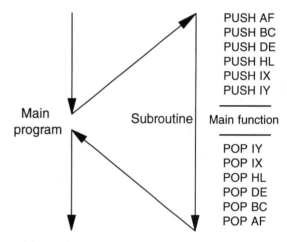

Fig. 3.8   Z80 subroutine with push and pop instructions.

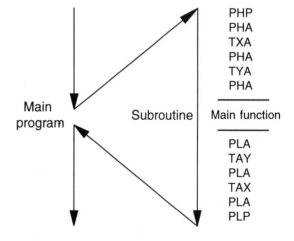

Fig. 3.9   6502 subroutine with push and pull instructions.

Interrupt Requests

An *interrupt request* is an externally generated signal which is applied to a particular pin on the microprocessor. Its purpose is to ask the microprocessor to suspend execution of the main program, then perform some action, before resuming its previous task.

Consider a simple analogy where the telephone rings as you are reading a book. You have just been *interrupted*! Assuming that you decide to answer the telephone, you might put down the book (leaving the book open or marking the page), before answering the telephone. On completion of the conversation, you are then able to resume reading the book at the point reached.

In fact more complex situations than this may easily arise. Imagine for example that there is a knock at the door as you answer the telephone. You now have two interrupts to deal with at the same time! A decision about the relative importance or *priority* of the two interrupts must be made before proceeding – you might ask the telephone caller to wait a moment while you answer the door. Once you have dealt with the personal caller, the telephone conversation may be resumed, after which you can continue reading the book.

An interrupt to a microprocessor may involve a sensor producing an alarm condition, or a communication interface reporting that incoming data has been received. Despite the different nature of the interrupt signal, microprocessors deal with interrupts in a very similar way.

When an interrupt request occurs, the microprocessor firstly decides whether to acknowledge the request, or to ignore it and continue with the main program. This decision depends on the relative importance of the interrupt and the main program. If the microprocessor does not want to be disturbed (because the main program is too important to be interrupted), it may decide to disable incoming interrupt requests.

Many microprocessors have more than one interrupt request pin, one of which has higher priority than the other. In general, the higher priority interrupt request cannot be ignored.

The *interrupt request* (IRQ or INT) signal may be enabled or disabled under software control, while the *non-maskable interrupt* (NMI) signal cannot be ignored by the microprocessor. Not surprisingly, non-maskable interrupts are generally reserved for external events of great importance, where a fast response is vital.

Once the decision to accept the interrupt request is made, the status of the microprocessor must be saved, so that the main program may later be resumed at the point reached. This is achieved by pushing the content of the program counter register onto the stack, which is performed automatically by the microprocessor. A jump is then made to a special program called an *interrupt service routine* (ISR), whose responsibility is to deal with the interrupt.

The ISR is very similar to a subroutine and is terminated by a *return from interrupt* instruction which causes the old program counter to be restored from the stack (allowing the main program to resume). The ISR may also alter other registers used by the main program, so these must also be saved and later restored. As with subroutines, a series of push instructions are used at the start of the ISR to save any data registers used during the servicing of the interrupt. These are later restored to their original state using a sequence of pop (or pull) instructions, prior to returning from the ISR. Figure 3.10 illustrates this sequence of events.

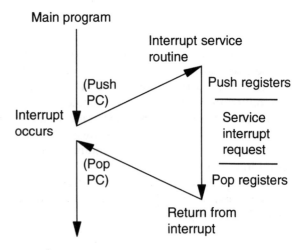

Fig. 3.10   Interrupt operation.

Interrupts and the Z80

The Z80 provides two interrupt request pins which may be used by external devices requiring attention from the microprocessor. As in the previous discussion, the *non-maskable interrupt* ($\overline{NMI}$) signal cannot be ignored, while the less important *interrupt request* ($\overline{INT}$) signal can be enabled or disabled under software control. Both signals are active low, as indicated by the inversion bar above their names. Table 3.5 summarises the action of each signal.

Pin	Activated by	Comments
$\overline{NMI}$	Negative edge	Cannot be disabled.
$\overline{INT}$	Active low	Enabled using the EI (enable interrupt) instruction. Disabled by the DI (disable interrupt) instruction.

Table 3.5 Z80 interrupt signals.

Note that the Z80 also has a *bus request* ($\overline{BUSRQ}$) input, which is used in *direct memory access (DMA)* applications. All descriptions of interrupt operation assume that $\overline{BUSRQ}$ is inactive (logic 1).

As can be seen from table 3.5, maskable interrupts may be enabled or disabled by the *enable interrupt (EI)* and *disable interrupt (DI)* instructions. By default (following a reset condition), maskable interrupts are disabled, so they must be enabled by the programmer before use.

There are also subtle differences in the way that the two interrupt types are activated. If the $\overline{NMI}$ input is pulled to a low logic level and held there, only a single interrupt is generated (by the one to zero transition). The $\overline{INT}$ signal, due its level sensitive nature would generate a continuous stream of interrupt requests under the same conditions.

Whether maskable interrupts are enabled or disabled is controlled by the state of a flip-flop called *IFF$_1$ (interrupt flip flop one)*. This bistable is set or cleared by the EI and DI instructions.

During non-maskable interrupts, IFF_1 is automatically cleared, to disable any possible interrupt requests. On completion of the interrupt service routine, the previous state of IFF_1 is restored. Thus, if interrupt requests were enabled before the non-maskable interrupt occurred, they are enabled again on completion of the ISR.

A second flip flop called IFF_2 is used as a temporary store for the state of IFF_1 during the non-maskable interrupt service routine. Table 3.6 summarises the behaviour of IFF_1 and IFF_2.

Action	IFF_1	IFF_2	Operation
Default	0	0	$\overline{INT}$ disabled
DI	0	0	$\overline{INT}$ disabled
EI	1	1	$\overline{INT}$ enabled
$\overline{NMI}$ begins	0	IFF_1	IFF_1 to IFF_2
$\overline{NMI}$ ends	IFF_2	●	IFF_2 to IFF_1

Table 3.6 IFF_1 and IFF_2 operation.

A consequence of the higher priority given to the $\overline{NMI}$ interrupt is that a partially completed maskable interrupt service routine may be interrupted, if an $\overline{NMI}$ occurs before its completion. This is illustrated by figure 3.11.

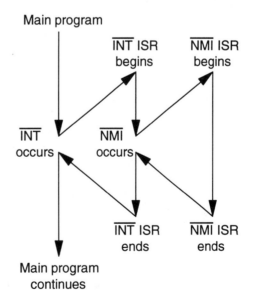

Fig. 3.11 $\overline{NMI}$ priority over $\overline{INT}$.

The detailed operation of each Z80 interrupt type will now be examined.

When a non-maskable interrupt occurs, the program counter register is pushed onto the stack and the current state of IFF_1 is stored in IFF_2. IFF_1 is cleared, thus disabling maskable interrupts. A jump is then made to address 0066_{16}, which should contain the non-maskable interrupt service routine (or a jump to the actual ISR).

On completion of the ISR, a *return from non-maskable interrupt (RETN)* instruction causes the previous program counter value to be popped from the stack and the state of IFF_2 to be copied back to IFF_1.

It is important to realise that it is the programmer's responsibility to ensure that a suitable ISR exists at memory location 0066_{16}, if non-maskable interrupts are to be used. The ISR should also ensure that the content of any registers used is unaltered, by the use of PUSH and POP instructions. Listing 3.1 shows a simple test program and non-maskable interrupt service routine.

```
        ORG 0066H    ;NMI ISR address
        PUSH AF      ;Save A + flags
        IN A,(DRB)   ;Read last count
        INC A        ;Count up on port B
        OUT (DRB),A  ;Send new value
        POP AF       ;Restore A + flags
        RETN         ;Return from NMI

        ORG 4000H    ;Main program start
DRA:    EQU 0        ;Z80 PIO port
CRA:    EQU 2        ;locations
DRB:    EQU 1
CRB:    EQU 3
        LD SP,5000H  ;Stack = top of RAM
        LD A, 0FH    ;Byte output
        OUT (CRA),A  ;on port A
        OUT (CRB),A  ;and port B
        LD A,0       ;Set ports A
        OUT (DRB),A  ;and B to zero
LP:     OUT (DRA),A
        INC A        ;Count up on port A
        JP LP
```

Listing 3.1　Z80 NMI test program.

The operation of this program may be tested by connecting a low frequency pulse waveform to the $\overline{\text{NMI}}$ input, while observing the output on ports A and B. Port A is incremented at high speed by the main program (use an oscilloscope to observe this). Port B is incremented much more slowly, on each high to low transition of the pulse waveform.

In practice, slight alteration may be needed to enable this program to work on a particular Z80 system. The main problem is that immediately following a reset condition, the Z80 begins executing instructions at address 0000_{16} (the program counter contains zero after a reset). ROM-based resident software is likely to be placed in this area of the memory map, to ensure it is executed after the reset condition has ended. Several solutions are possible.

- Use an EPROM programmer to create your own EPROM containing the non-maskable interrupt service routine. (The Z80's reset vector must also be catered for.)
- In some computers, a jump instruction (op. code $C3_{16}$) is placed at address 0066_{16}. This jump instruction causes the $\overline{\text{NMI}}$ vector to be redirected to an area of memory which is RAM. In this case, read the computer manual for details of creating your own $\overline{\text{NMI}}$ service routine.

Maskable interrupts, activated by the $\overline{\text{INT}}$ signal, are rather more complex than non-maskable interrupts. This is because the Z80 is capable of operating in three different interrupt modes, as selected by the *select interrupt mode* instruction. Table 3.7 lists the available interrupt types and the instructions used to select them.

Instruction	Comments
EI	Enable maskable interrupts
DI	Disable maskable interrupts
IM 0	Select interrupt mode zero
IM 1	Select interrupt mode one
IM 2	Select interrupt mode two

Table 3.7　Z80 maskable interrupt instructions.

Practical Exercise

Non-maskable Interrupts

Method

Use an assembler to enter the appropriate program into memory (adjusting ORG addresses and interrupt vectors as necessary to suit the computer memory map).

Connect an 8 bit input/output monitor unit to each port, to allow the value on each port to be examined.

Connect the pulse generator to the $\overline{\text{NMI}}$ pin and set the pulse repetition frequency to 1 cycle per second.

Execute the main program.

Equipment required

Z80 or 6502 based microcomputer
PIO IC with two 8 bit ports
Pulse generator
Oscilloscope (for frequency measurements)
Two 8 bit input/output monitor units (LEDs)

Conclusions

Measure the speed of count on ports A and B. Steadily increase the pulse generator frequency, while recording the speed of count on each port. Comment on your findings.

6502 Test Program

```
        ORG $FFFA      ;Set NMI vector
        DEFB $00,$48   ;to $4800

        ORG $4000      ;Main program
DRA:    EQU $9000      ;6502 port
DDRA:   EQU $9001      ;addresses
DRB:    EQU $9002
DDRB:   EQU $9003
        LDX #$FF        ;Initialise
        TXS             ;stack pointer
        LDA #$FF        ;All bits output
        STA DDRA        ;on ports A
        STA DDRB        ;and B
        LDA #0
        STA DRB         ;Set ports A and
        STA DRA         ;B to zero
LP:     INC DRA         ;Count up on A
        JMP LP

        ORG $4800       ;NMI ISR
        INC DRB         ;Count up on B
        RTI             ;Return from NMI
```

Z80 Test Program

```
        ORG 0066H      ;NMI ISR address
        PUSH AF        ;Save A + flags
        IN A,(DRB)     ;Read last count
        INC A          ;Count up on B
        OUT (DRB),A    ;Send new value
        POP AF         ;Restore A + flags
        RETN           ;Return from NMI

        ORG 4000H      ;Main program
DRA:    EQU 0          ;Z80 PIO port
CRA:    EQU 2          ;locations
DRB:    EQU 1
CRB:    EQU 3
        LD SP,5000H    ;Top of RAM
        LD A, 0FH      ;Byte output
        OUT (CRA),A    ;on port A
        OUT (CRB),A    ;and port B
        LD A,0         ;Set ports A
        OUT (DRB),A    ;and B to zero
LP:     OUT (DRA),A
        INC A          ;Count up on A
        JP LP
```

Interrupt mode zero requires the interrupting device to place a valid Z80 op. code on the data bus, in response to a Z80 *interrupt acknowledge cycle*. This mode, which is not commonly used, provides compatibility with the earlier Intel 8080 microprocessor.

An interrupt acknowledge cycle is generated in response to an interrupt request, following the completion of the current instruction, provided that maskable interrupts are enabled. This special machine cycle is similar to an op. code fetch, with the exception that $\overline{\text{IORQ}}$ and M_1 are active, rather than $\overline{\text{MREQ}}$ and $\overline{\text{M}_1}$. A simple circuit, capable of detecting a Z80 interrupt acknowledge cycle is shown in figure 3.12.

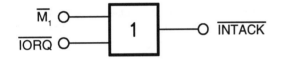

Fig. 3.12   Z80 interrupt acknowledge circuit.

Once the interrupt acknowledge pin goes low, the interrupting device places an op. code on the data bus. This instruction is then executed by the microprocessor.

The most commonly used instruction type is the *restart* (*RST*) instruction, which provides a single byte subroutine call. The RST instruction is extremely compact, compared to the CALL instruction, which requires three bytes of memory. The disadvantage is that the RST instruction can only jump to one of eight possible addresses in page zero, as shown in table 3.8.

Address	Instruction	Comments
0000	RST 0H	Reset address
0008	RST 08H	
0010	RST 10H	
0018	RST 18H	
0020	RST 20H	
0028	RST 28H	
0030	RST 30H	
0038	RST 38H	Mode 1 address

Table 3.8   RST instructions and addresses.

Due to the small space between each RST entry point, it is normal practice to place the interrupt service routine elsewhere in memory. A jump instruction is then placed at the RST address, causing the microprocessor to branch to the actual ISR. In addition, since the RST instruction is actually a subroutine call, the ISR must be terminated using the normal return from subroutine instruction (RET), rather than a return from interrupt.

Maskable interrupts are automatically disabled once the interrupt has been acknowledged. This is necessary because of the level sensitive nature of the interrupt signal, and prevents repeated calls to the ISR.

The last-but-one instruction in the ISR is normally the enable interrupt (EI) instruction. The action of this instruction is delayed by one instruction, in order to allow a return to the main program before any further interrupts are acknowledged. This delayed action is vital, in order to prevent an unwanted build-up of data on the stack, due to a continuous stream of interrupt requests. Listing 3.2 shows the structure of a typical mode 0 interrupt driven program, which uses the RST 20H vector.

```
        ORG 0020H   ;RST 20H address
        JP ISR      ;Go to ISR

        ORG 4000H   ;Main program
DRA:    EQU 0       ;Z80 PIO port
CRA:    EQU 2       ;locations
        LD SP,5000H ;Stack = top of RAM
        EI          ;Enable interrupts
        IM 0        ;Mode 0
        LD A, 0FH   ;Byte output
        OUT (CRA),A ;on port A
        LD A,0      ;All LEDs off -
LP:     OUT (DRA),A ;when in main loop
        JP LP
ISR:    PUSH AF     ;Save A + flags
        LD A,0FFH   ;All LEDs on -
        OUT (DRA),A ;when in ISR
        POP AF      ;Restore A + flags
        EI          ;Enable interrupts
        RET         ;Return from INT
```

Listing 3.2   Demonstration mode 0 program.

Practical Exercise

Mode 0 Interrupts (Z80 only)

Introduction

The circuit shown will output an op. code (as defined by the switches), in response to a Z80 interrupt acknowledge cycle. It may therefore be used to test interrupt service routines, based on mode 0 operation.

Method

Construct the circuit and connect it to the indicated microprocessor signals. Set the switches so that they represent the op. code of one of the RST instructions shown in the table opposite.

Write a suitable main program and interrupt service routine, so that some visible action is performed when the INT pin goes low.

(Momentarily connect the INT pin to 0 volts to cause an interrupt request.)

Equipment required

Z80 based microcomputer
8 10 kΩ resistors
8 SPST switches (on/off)
74LS244 (octal buffer with tri-state outputs)
74LS32 (quad 2 input OR gates)
Prototyping board

Restart Op. codes

RST 00H	$1100\ 0111_2$
RST 08H	$1100\ 1111_2$
RST 10H	$1101\ 0111_2$
RST 18H	$1101\ 1111_2$
RST 20H	$1110\ 0111_2$
RST 28H	$1110\ 1111_2$
RST 30H	$1111\ 0111_2$
RST 38H	$1111\ 1111_2$

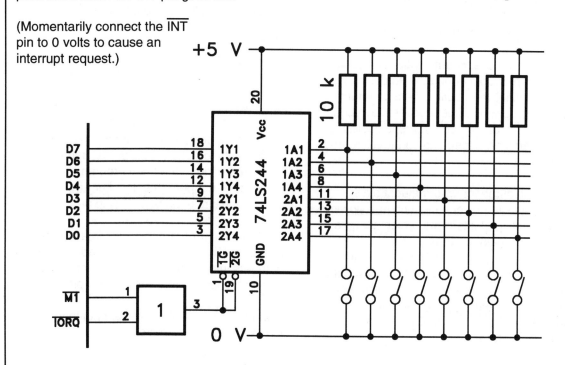

The second method of dealing with maskable interrupts is called *interrupt mode one*. This mode is selected by the IM 1 instruction (remembering also to enable maskable interrupts).

As with non-maskable interrupts, a mode one interrupt causes a jump to a user defined interrupt service routine, at a fixed address in memory.

Following an interrupt request, the program counter is pushed onto the stack. The microprocessor then jumps to addresses 0038_{16} (which is also the address used by the RST 38H instruction).

Once again, this area of memory is likely to be in ROM. In many systems, a jump instruction is placed at this address, causing the mode 1 interrupt to be re-directed to an area of RAM.

For example, if addresses 0038_{16}, 0039_{16} and $003A_{16}$ contains $C3_{16}$, 00_{16} and 40_{16} respectively, the microprocessor would jump to an ISR at address 4000_{16}, following a mode one interrupt.

The mode one ISR should be terminated with the return from interrupt RETI instruction, rather than with the RET instruction used for mode 0. (The use of RET with mode zero assumed that the external peripheral placed the op. code of an RST instruction on the data bus in response to a Z80 interrupt acknowledge cycle.)

Listing 3.3 shows the structure of a typical Z80 program which uses interrupt mode one.

Practical Exercise   Program Single Step Facility

Introduction

If the maskable interrupt pin is connected to logic 0, and held there, an interrupt service routine will be repeatedly called from the main program (if interrupts have been enabled).

Most microprocessors automatically disable maskable interrupts during the actual interrupt service routine, re-enabling them on return to the main program. (This process may occur automatically, or may be the responsibility of the programmer, depending on the microprocessor type.)

The testing of the interrupt request pin is normally delayed by one instruction, following the re-enabling of interrupts, which allows a return to the main program before the next interrupt occurs. This has the effect that one instruction from the main program is executed each time the ISR is called. This microprocessor characteristic may be used, in conjunction with a suitable ISR, to *single step* the main program.

Method

For the microprocessor of your choice, write a maskable interrupt service routine which performs the following functions.

- displays the content of all data registers.
- displays the current value of the program counter register. (This value was pushed onto the stack on entry to the ISR – be careful to correct the stack pointer after reading this value!)
- waits for a key to be pressed, before returning to the main program.
- does not permanently alter any of the registers used by the main program.

Write a main program that runs in a continuous loop. The program should load various numbers into each data register, and must begin by enabling maskable interrupts.

Connect the maskable interrupt pin to logic 0, to test the single step facility.

```
        ORG  0038H   ;RST 38H address
        JP ISR       ;Go to ISR

        ORG  4000H   ;Main program
DRA:  EQU 0          ;Z80 PIO port
CRA:  EQU 2          ;locations
      LD SP,5000H    ;Stack = top of RAM
      EI             ;Enable interrupts
      IM 1           ;Mode 1
      LD A, 0FH      ;Byte output
      OUT (CRA),A    ;on port A
      LD A,0         ;All LEDs off -
LP:   OUT (DRA),A    ;when in main loop
      JP LP

ISR:  PUSH AF        ;Save A + flags
      LD A,0FFH      ;All LEDs on -
      OUT (DRA),A    ;when in ISR
      POP AF         ;Restore A + flags
      EI             ;Enable interrupts
      RETI           ;Return from INT
```

Listing 3.3 Demonstration mode 1 program.

Compare the above program with listing 3.2 to see the necessary changes for mode one operation. In particular, notice the use of a jump instruction to move the ISR to an area of RAM and the *return from interrupt request* instruction (RETI) at the end of the ISR.

Mode two operation is the most powerful method of dealing with maskable interrupts, as it permits multiple interrupt sources to be connected to the INT pin. Each external device is allowed to have its own interrupt service routine, located at a unique address. A mechanism is provided which allows the Z80 to identify rapidly the interrupting device and then execute the appropriate ISR

In fact, the microprocessor uses a look-up table in memory to hold the actual address of each ISR. Each 16 bit entry in this table is said to be an *indirect pointer* to the ISR.

Mode two operation is intended mainly for use with members of the Z80 peripheral family, such as the *Z80-PIO (parallel input/output)* and *Z80-CTC (counter/timer)* devices, which are considered in detail in the next chapter.

In response to a mode two interrupt request, the Z80 performs an interrupt acknowledge cycle. The interrupting device then places an 8 bit number onto the data bus, which represents the lower half of a vector, which points to an entry in a look-up table. The externally supplied byte is combined with the content of the I (interrupt) register to form a 16 bit vector. This process is shown in figure 3.13.

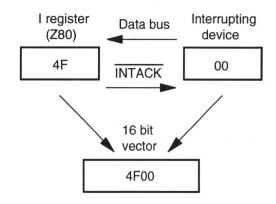

Fig. 3.13 Formation of a mode two interrupt vector.

The resulting 16 bit number refers to an indirect pointer to the ISR, which is stored 'low byte first'. This pointer is the actual address of the ISR. As will be seen in the next chapter, the data byte supplied by the external device is always an even number, ensuring that two bytes are allocated for each pointer.

Mode two operation requires that the I register should be initialised to the most significant byte of the ISR indirect pointer address, as part of the initialisation process. This is achieved by the LD I,A instruction, which copies the content of the accumulator into the I register. Members of the Z80 peripheral family must also be configured to output the correct interrupt vector during an interrupt acknowledge cycle. Generally this is achieved by storing the appropriate byte into a special purpose register inside the peripheral device. Once again, this task is performed during the initialisation section of the program, and is the programmer's responsibility.

Practical Exercise   Mode 2 Interrupts (Z80 only)

Introduction

The circuit shown will output the least significant byte of an interrupt vector (as defined
by the switches), in response to a Z80
interrupt acknowledge cycle. It may therefore be used to test interrupt service
routines, based on mode 2 operation.

Equipment required

Z80 based microcomputer
8 10 kΩ resistors
8 SPST switches (on/off)
74LS244 (octal buffer with tri-state outputs)
74LS32 (quad 2 input OR gates)
Prototyping board

Method

Construct the circuit and connect it to the
indicated microprocessor signals. Set the
switches so that they represent the least
significant byte of the interrupt vector
address.

Write a suitable main program and interrupt
service routine, so that some visible action is
performed when the INT pin
goes low.

Program Initialisation

```
ORG 4000H    ;Start of main
LD SP,5000H  ;Stack = top of RAM
EI           ;Enable interrupts
IM 2         ;Interrupt mode 2
LD A,48H     ;MSB of vector
LD I,A       ;to I register

ORG 4800H    ;ISR vector table
             ;vector 1 etc.
```

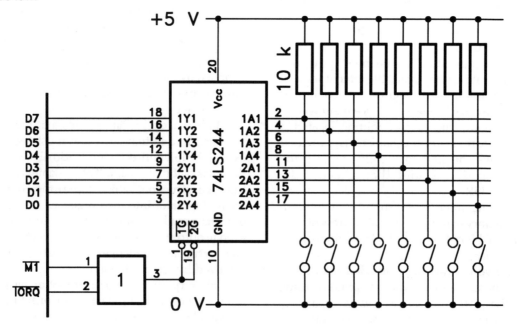

The general structure of a typical Z80 program using interrupt mode two is shown in listing 3.4.

```
ORG 4000H    ;Main program
LD SP,5000H  ;Stack = top of RAM
EI           ;Enable interrupts
IM 2         ;Mode 2
LD A,44H     ;MSB of ISR vector
LD I,A       ;table address
             ;Put device init-
             ;ialisations here,
             ;+ the main program
HALT


ORG 4400H    ;ISR vector table
DEFW 4800H   ;ISR1 pointer
DEFW 4840H   ;ISR2 pointer
DEFW 4880H   ;ISR3 pointer


ORG 4800H    ;ISR1 entry point
             ;Push registers
             ;ISR1 function here
             ;Pop registers
EI           ;Enable interrupts
RETI         ;Return from INT


ORG 4840H    ;ISR2 entry point
             ;Push registers
             ;ISR2 function here
             ;Pop registers
EI           ;Enable interrupts
RETI         ;Return from INT


ORG 4880H    ;ISR3 entry point
             ;Push registers
             ;ISR3 function here
             ;Pop registers
EI           ;Enable interrupts
RETI         ;Return from INT
```

Listing 3.4 Structure of a mode 2 program.

The three ISRs in the above program are each associated with a particular external device.

It is assumed here that device one outputs 00_{16} onto the data bus in response to an interrupt acknowledge cycle, while devices two and three output 02_{16} and 04_{16} respectively.

As mentioned previously, any data registers used by the main program must not be altered by the ISR. In most microprocessors, including the Z80, this may be achieved by pushing registers onto the stack at the start of the ISR and later popping them from the stack (in reverse order), before returning to the main program.

It should also be noted that the Z80 has an alternate register set, and a set of register exchange instructions, which may be used for rapid *context switching*. In fact, the time taken to *exchange* all Z80 registers is considerably less than that required for the equivalent push or pop instructions! Listing 3.5 shows a simple ISR which leaves all Z80 registers unaltered.

```
ISR: EXX         ;ISR entry point
     EX AF,AF'   ;Swap registers
                 ;Put ISR function
                 ;here
     EXX         ;Restore registers
     EX AF,AF'   ;used by MAIN prog.
     RETI        ;- or RET or RETN
```

Listing 3.5 ISR which uses register swapping.

The above method is ideal in applications where only a single interrupt may occur at any time. If there is any possibility of two (or more) interrupts being simultaneously active, then this technique should be used with extreme caution.

Consider an example where a maskable interrupt occurs, causing a jump to the appropriate ISR. Assuming that register swapping is used, the alternate register set becomes active during the ISR. If a non maskable interrupt now occurs, before completion of the first ISR, this will cause the maskable interrupt ISR to be temporarily suspended. If this second ISR also uses register swapping, the main register set will once again become active. The second ISR may now corrupt registers being used by the main program, causing the main program to malfunction, when it later resumes.

This type of programming error can cause software faults which are obscure and difficult to trace. If in doubt, use push and pop instructions, rather than exchanging registers!

Interrupts and the 6502

The 6502 provides two interrupt request pins which may be used by external devices requiring attention from the microprocessor. As mentioned at the start of this chapter, the *non-maskable interrupt* ($\overline{NMI}$) signal cannot be ignored, while the less important *interrupt request* ($\overline{IRQ}$) signal can be enabled or disabled under software control. Both signals are active low, as indicated by the inversion bar above their names. Table 3.9 summarises the action of each signal.

Pin	Activated by	Comments
$\overline{NMI}$	Negative edge	Cannot be disabled.
$\overline{IRQ}$	Active low	Enabled using the CLI (clear interrupt flag) instruction. Disabled by the SEI (set interrupt flag) instruction.

Table 3.9 6502 interrupt signals.

As can be seen from table 3.9, maskable interrupts may be enabled or disabled by either clearing or setting the interrupt mask flag. This is achieved using the CLI and SEI instructions, which directly affect the state of the processor status register. By default (following a reset condition), maskable interrupts are disabled, so they must be enabled by the programmer before use. Non-maskable interrupts, on the other hand, cannot be disabled. Figure 3.14 shows the layout of the 6502 processor status register.

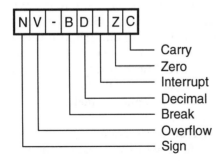

Fig. 3.14 6502 processor status register.

There are also subtle differences in the way that the two interrupt types are activated. If the $\overline{NMI}$ input is pulled to a low logic level and held there, only a single interrupt is generated (by the one to zero transition). The $\overline{IRQ}$ signal, due its level sensitive nature, would generate a continuous stream of interrupt requests under the same conditions.

A consequence of the higher priority given to the $\overline{NMI}$ interrupt is that a partially completed maskable interrupt service routine may be interrupted, if an $\overline{NMI}$ occurs before its completion. This is illustrated by figure 3.15.

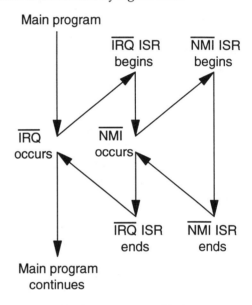

Fig. 3.15 $\overline{NMI}$ priority over $\overline{IRQ}$.

When an interrupt is accepted by the microprocessor, the current instruction is completed, after which the program counter and the processor status register are pushed onto the stack. The interrupt mask (I) flag is then cleared, causing maskable interrupts to be disabled.

Pushing the I flag onto the stack enables the previous maskable interrupt status (enabled or disabled) to be later restored. This occurs automatically when the P register is pulled from the stack, on completion of the ISR. (This technique is also effective during nested ISR calls, of the type shown in the above figure.)

The detailed operation of each 6502 interrupt type will now be considered.

When a *non-maskable interrupt* occurs, the program counter register is pushed onto the stack, followed by the processor status register. The I flag is then cleared, disabling maskable interrupts. The address of the ISR is then obtained by reading the content of addresses $FFFA_{16}$ and $FFFB_{16}$. These two memory locations contain a *vector* (stored low byte first), which points to the interrupt service routine. Other vectors (the maskable interrupt vector and the microprocessor *reset vector*), are also stored in this area of memory, as shown in figure 3.16.

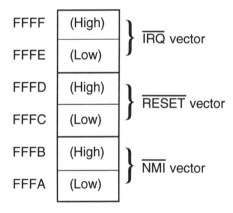

Fig. 3.16 6502 interrupt and reset vectors.

On completion of the ISR, a *return from interrupt* (RTI) instruction causes the previous program counter value to be popped from the stack and the flags register to be restored to its previous state.

It is important to realise that it is the programmer's responsibility to ensure that a suitable ISR exists at the address pointed to by the NMI vector. The ISR should also ensure that the content of any registers used is unaltered, by the use of PUSH and PULL instructions. Listing 3.6 shows a simple test program and non-maskable interrupt service routine.

```
ORG $FFFA     ;Set NMI vector
DEFB $00,$48  ;to $4800

ORG $4000     ;Main program
```

```
DRA:    EQU $9000    ;6502 port
DDRA:   EQU $9001    ;addresses
DRB:    EQU $9002
DDRB:   EQU $9003
        LDX #$FF     ;Initialise
        TXS          ;stack pointer
        LDA #$FF     ;All bits output
        STA DDRA     ;on ports A
        STA DDRB     ;and B
        LDA #0
        STA DRB      ;Set ports A and
        STA DRA      ;B to zero
LP:     INC DRA      ;Count up on A
        JMP LP

        ORG $4800    ;NMI ISR
        INC DRB      ;Count up on B
        RTI          ;Return from NMI
```

Listing 3.6 6502 NMI test program.

The operation of this program may be tested by connecting a low frequency pulse waveform to the NMI input, while observing the output on ports A and B. Port A is incremented at high speed by the main program (use an oscilloscope to observe this). Port B is incremented much more slowly, on each high to low transition of the pulse waveform.

In practice, slight alteration may be needed to enable this program to work on a particular 6502 system. The main problem is that immediately following a reset condition, the 6502 begins executing instructions at the vector pointed to by addresses $FFFC_{16}$ and $FFFD_{16}$ (so the program counter is loaded with this vector following a reset). This area of memory is likely to be ROM, to ensure that resident software is executed after the reset condition has ended. Several solutions are possible.

● Create your own EPROM containing a suitable vector for your non-maskable interrupt service routine.
● In some computers, the NMI vector is preset to point to an area of RAM. Read your computer manual for further details.

The structure of a typical 6502 interrupt driven program, based on maskable interrupts is very similar to the $\overline{\text{NMI}}$ based program considered earlier. The main differences are the level sensitive nature of the $\overline{\text{IRQ}}$ interrupt, the need to enable interrupts using the CLI instruction, and the use of memory locations FFFE_{16} and FFFF_{16} to hold the ISR vector.

To alter listing 3.6 for $\overline{\text{IRQ}}$ based operation, change the first ORG statement from \$FFFA to \$FFFE and insert a CLI instruction in the initialisation section. Notice that the behaviour of the program is significantly altered, due to the level sensitive nature of the $\overline{\text{IRQ}}$ input. Whenever $\overline{\text{IRQ}}$ is high, the main program runs at full speed. When $\overline{\text{IRQ}}$ is low, the ISR is repeatedly called.

An additional complication with maskable interrupts is that an interrupt can be generated in hardware, by a low level on the $\overline{\text{IRQ}}$ input, or in software by the 6502 *break* (*BRK*) instruction.

The BRK instruction effectively provides a *software interrupt* facility, which is often used for program debugging purposes (see the practical exercise below).

When writing an interrupt service routine for use with maskable interrupts, some way must be found to determine whether the interrupt was generated by external hardware, or by software. This may be achieved by examining the B (break) flag in the processor status register. The BRK instruction causes this flag to be set prior to pushing the P register onto the stack, while a hardware interrupt clears the B flag. Listing 3.7 shows a short routine which tests the state of the B flag, and branches accordingly.

```
TEST:   PLA         ;Get P register
        PHA         ;Correct the stack
        AND #$10    ;Mask unwanted bits
        BNE BK      ;Go to BRK handler
        ...         ;Hardware interrupt
```

Listing 3.7 Break flag test routine.

Practical Exercise   Breakpoints

Introduction

Breakpoints are commonly used program debugging tools, normally found in association with monitor software.

If a breakpoint is inserted in a program, the main program will execute normally until the breakpoint is reached, after which control is passed back to the monitor program. The user may then study the contents of any data registers, flags or memory locations which are relevant to the operation of the program. Further testing of the software is then possible by single stepping the program from the point reached. (Note that the method shown here for creating breakpoints is only effective for RAM based programs.)

Method

Enter a test (machine code) program into RAM.

Replace the op. code byte of the instruction to be tested with a suitable single byte instruction, which will cause a branch to a subroutine or interrupt service routine (use the 6502 BRK or the Z80 RST xxH instruction). Make a note of the address and its original content.

Write a program that displays the current state of the microprocessor, and then returns control to the monitor when the breakpoint is encountered.

Execute the test program, ensuring that control is passed back to the monitor program when the RST or BRK instruction is reached. Finally replace the breakpoint with the original data value.

Practical Exercise   Burglar Alarm

Introduction

If one or more of the inputs (zone 1 – zone 4) gives a logic 1 signal, this will generate an interrupt. The interrupt service routine which is called following an alarm should indicate the source of the alarm, as well as producing an audible signal. After a predetermined period, the alarm should be cancelled and testing of the inputs resumed.

Method

Connect the circuit shown to a suitable microcomputer (which must have an unused maskable interrupt pin and a parallel input/output port IC).

The ISR should cause the appropriate LED(s) to be illuminated, thus indicating the source of the alarm. This may be achieved by reading port A (mask off the most significant nibble), and then outputting this value to port B.

A continuous tone should also be generated by producing a square wave signal on the most significant bit of port B. (A simple way to toggle the most significant bit, while leaving other bits unaltered is to exclusive OR the port value with 80_{16}.) The frequency of oscillation should be within the audio frequency range (1 kHz for example), and should be matched to the transducer's resonant frequency for maximum volume.

After a suitable period (10 seconds is ample for test purposes), the main program should resume. (All LEDs should be out and the buzzer cancelled.)

Circuit Diagram

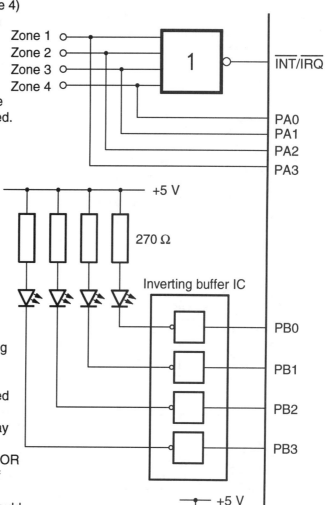

Interrupts and Polling

When external devices are being dealt with, which may from time to time request attention, interrupts are not the only solution. An alternative approach is to regularly test each device to check its status. Those devices which require attention are 'serviced', before moving on to test other devices in sequence. This approach, which is commonly known as *polling*, is illustrated by the flowchart of figure 3.17.

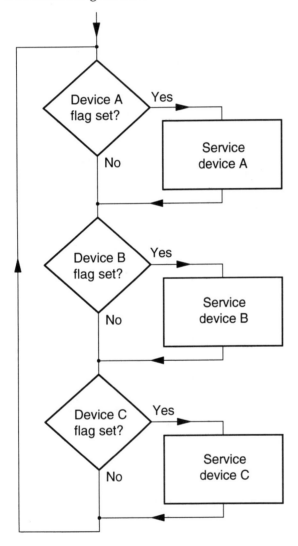

Fig. 3.17   Testing external devices by polling.

In practice, the 'flag' used to indicate the status of a particular peripheral device might be a single bit of an input port, as shown in figure 3.18 below.

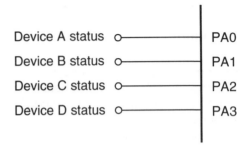

Fig. 3.18   Status flags wired to an input port.

The test program would read the port and then mask off all unwanted bits to determine the status of each flag in sequence.

The polled approach seen here may lead to slight simplification in external hardware, compared with an interrupt driven system. Generally this slight advantage is outweighed by the slow (and sometimes unpredictable) response time provided by polled systems.

In a typical polling situation, each flag is tested many times when service is not required. This is wasteful in terms of microprocessor time, particularly if the microprocessor is required to perform other tasks, as well. One solution to this problem is to test the flags less frequently, but this may lead to unacceptable delays between a device requesting and receiving attention.

Another problem with polling is that a device may request attention while other devices are being serviced. In certain situations there may be a significant delay before the device is serviced. This may make a polled system unsuitable for certain time critical applications.

It should also be noted that interrupt driven systems, although offering improved response times and the appearance of performing multiple simultaneous tasks (*multitasking*), are not without problems. The performance of an interrupt driven system gradually degrades as the frequency of interrupts increases, until all of the microprocessor's time is spent dealing with interrupts. Rapid non-maskable interrupts may also lead to stack overflow and catastrophic failure of the system!

Practical Exercise	Interrupt Driven Keyboard Matrix

Introduction

The interfacing of a keyboard arranged as a matrix of switches was considered in the first volume. At this time a *polled* approach was used, based on repeatedly *scanning* the keyboard.

The addition of a single logic gate allows an interrupt to be generated if a key is pressed, thus allowing the microprocessor to perform other tasks, while the keyboard is inactive.

Method

Write a main program which outputs 00_{16} to port A, as part of the main program's initialisation. This main program should also enable maskable interrupts and then perform some other task in a continuous loop.

If no keys are pressed, all inputs to port B will read as logic 1, due to the action of the pull-up resistors. The output of the AND gate will also be logic 1.

When a key is pressed, that column will be pulled to logic 0, due to the connection made to port A (both bits of port A are at logic 0). The AND gate will then generate an interrupt request.

Write an ISR that will scan the keyboard once only, to identify the pressed key. The (ASCII) character associated with the pressed key should then be printed on the screen.

Circuit Diagram

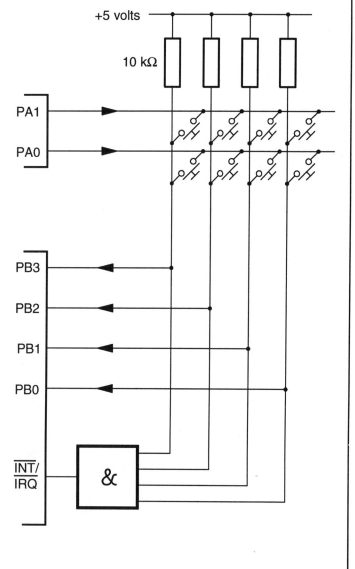

Interrupt Priority Systems

In a system that is connected to several possible interrupt sources, certain interrupts are likely to be more important (of higher *priority*) than others.

Consider the example of a microcomputer that is used to control a chemical plant. Inputs to the computer would be provided from sensors measuring parameters such as temperature, pressure, flow rate or level, while outputs would control *actuators* such as valves, heaters or motors.

Imagine the possible consequences of a power failure to the microcomputer. Motors and valves may be left in a dangerous state (running or open respectively), leading to an increase in temperature and a possible explosion! In such a situation it would be very important that the system should be shut down safely in the event of a power failure.

A circuit could be designed that would give advance warning of power loss by monitoring the mains voltage directly. This would allow the system to shut down safely before power supply capacitances, or emergency backup batteries were discharged. Such an alarm signal could then be connected to the non-maskable interrupt pin, leading to a rapid response, regardless of the microprocessor state prior to the interrupt.

Not all cases of interrupt priority are as dramatic as the example given above. However it is still useful to be able to arrange interrupt signals in ascending order of priority. This ensures that the microprocessor is able to devote most of its valuable time to dealing with important events, rather than ones which are not urgent.

A typical microprocessor, such as the 6502 or Z80, already has two levels of interrupt priority. These are the maskable and non-maskable interrupt inputs which have already been considered. If more than two levels of priority are required then some way must be found to connect multiple external interrupt signals to a single microprocessor pin. A simple solution is to use a logic gate, as shown in figure 3.19.

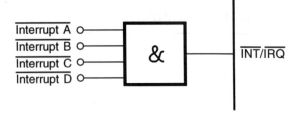

Fig. 3.19 Connection of multiple interrupts.

If one or more of the inputs goes to logic 0, the output of the AND gate becomes logic 0, thus generating an interrupt request. (The AND gate acts as a negative logic OR gate.) Alternatively, for positive logic inputs, a four input NOR gate could be used to generate the interrupt request.

Having generated an interrupt, the microprocessor now has the problem of determining which of the external devices generated the interrupt. One solution is to connect each interrupt request to an input port, as shown in fig. 3.20.

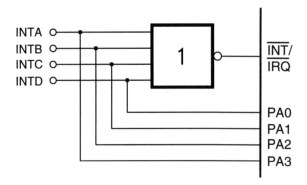

Fig. 3.20 Multiple interrupts with status flags.

On reception of an interrupt request, the microprocessor may now test (or poll) each of the interrupt flags in turn, using bit masking techniques. If a particular flag is set, its associated device is serviced by the microprocessor.

The order in which particular flags are tested is determined by the level of priority associated with each device. This method of prioritising interrupts is software intensive and hence is rather slow. It does have the advantage of greater flexibility than a purely hardware based solution, since interrupt priority can be altered by changing the software.

Practical Exercise | Interrupt Priority (Polled Approach)

Introduction

This exercise demonstrates the connection of four external (active high) interrupt signals to the microprocessor's maskable interrupt pin.

A polled approach is used to determine the identity of the interrupting device.

Method

Connect the circuit shown to the appropriate microprocessor signals.

Write a program and interrupt service routine that will print a single letter on the screen, each time an interrupt is generated. Use a different letter for each external device, to enable correct operation of the program to be easily determined.

Confirm that the program always prints the letter associated with the highest priority active input, even when more than one input is simultaneously at logic 1.

Circuit Diagram

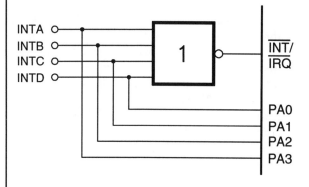

ISR Flowchart

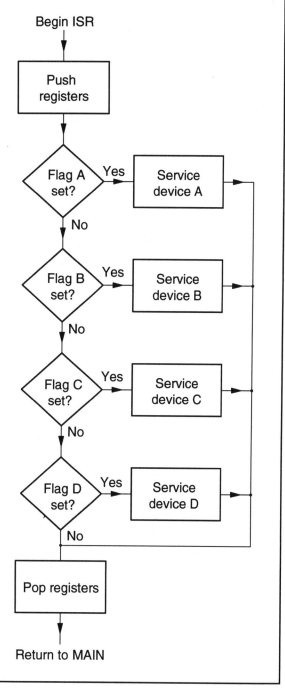

In the previous discussion, a separate logic gate was used to connect multiple interrupt requests to a single microprocessor pin. Certain types of IC (which are capable of generating interrupt requests), have special outputs which may be directly wired together. When connected, these wired outputs automatically combine to produce a logic function, eliminating the need for additional logic circuitry.

These outputs, which are commonly referred to as open collector (or open drain) circuits, are incapable of pulling their outputs to a high logic level. An external pull-up resistor is required for this purpose, as shown in figure 3.21.

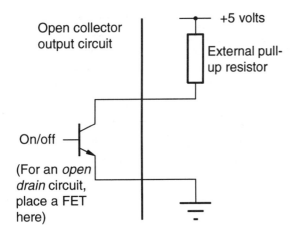

Fig. 3.21   Open collector output circuit.

With this type of output circuit, a high logic level is produced passively, via the pull-up resistor. The transistor turns on to actively pull the output to a low voltage.

Unlike 'normal' output circuits, it is possible to connect several open collector outputs together, using a single pull-up resistor to produce passively the logic 1 level. In this case the output voltage will be high if all output transistors are off. If one or more output transistors conducts, the output will be pulled to a low logic level.

This arrangement is commonly called a *wired-OR* circuit, since an active low (or negative logic) output is produced when one or more of the open collector transistors is switched on. Figure 3.22 illustrates the connection of several open collector gates to produce a single output.

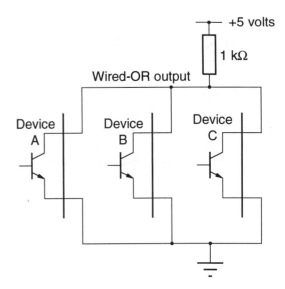

Fig. 3.22   Wired-OR connection of outputs.

To summarise, if the data sheets for a particular peripheral IC specify that the interrupt request output is open collector or open drain, this means that a pull-up resistor must be used for correct operation. It is also possible to wire together several such outputs, by sharing a single pull-up resistor. This wired-OR arrangement will generate an interrupt request if one or more of the output transistors is enabled.

In the above discussion, several possible sources of interrupt were combined logically to produce a single interrupt request. The microprocessor used a polled approach to identify the highest priority active device. A second method is to design hardware that automatically provides either the identification number of the highest priority interrupting device, or the actual address (the *interrupt vector*) of the ISR.

This second approach has the advantage that the microprocessor is able to respond to interrupts more quickly. However, there may be some loss of flexibility, since it more difficult to alter interrupt priorities which are fixed in hardware.

Figure 3.23 shows data sheets for the 74LS148 8 line to 3 line *priority encoder* IC, which is commonly used in hardware-based *interrupt priority systems*. (Reproduced by courtesy of Texas Instruments.)

TYPES SN74148 (TIM9907), SN74LS148
SN54148, SN54LS148
8-LINE TO 3-LINE PRIORITY ENCODERS

OCTOBER 1976 - REVISED DECEMBER 1983

- Encodes 8 Data Lines to 3-Line Binary (Octal)

- Applications Include:

  N-Bit Encoding
  Code Converters and Generators

- **Package Options Include Standard Plastic (N) and Ceramic (J) 300-mil Dual-In-Line Packages, Plastic Small Outline (D) and Ceramic Chip Carrier (FK) Package**

SN54148, SN54LS148...J PACKAGE
SN74148...N PACKAGE
SN74LS148....D OR N PACKAGE
(TOP VIEW)

'148, 'LS148

```
        ___
   4 [ 1   16 ] VCC
   5 [ 2    15 ] E0
   6 [ 3    14 ] GS
   7 [ 4    13 ] 3
  EI [ 5    12 ] 2
  A2 [ 6    11 ] 1
  A1 [ 7    10 ] 0
 GND [ 8     9 ] A0
```

SN54LS148 ... FK PACKAGE
(TOP VIEW

'LS148

```
          5  4  NC VCC E0
          3  2  1  20 19
     6 [ 4            18 ] GS
     7 [ 5            17 ] 3
    NC [ 6            16 ] NC
    EI [ 7            15 ] 2
    A2 [ 8            14 ] 1
          9  10 11 12 13
          A1 GND NC A0 0
```

NC - No internal connection

TYPE	TYPICAL DATA DELAY	TYPICAL POWER DISSIPATION
'148	10 ns	190 mW
'LS148	15 ns	60 mW

description

These TTL encoders feature priority encoding of the inputs to ensure that only the highest-order data line is encoded. The '148 and 'LS148 encode eight data lines to three-line (4-2-1) binary (octal). Cascading circuitry (enable input EI and enable output EO) has been provided to allow octal expansion without the need for external circuitry. For all types, data inputs and outputs are active at the low logic level. All inputs are buffered to represent one normalized Series 54/74 or 54LS/74LS load, respectively.

'148, 'LS148
FUNCTION TABLE

INPUTS									OUTPUTS				
EI	0	1	2	3	4	5	6	7	A2	A1	A0	GS	EO
H	X	X	X	X	X	X	X	X	H	H	H	H	H
L	H	H	H	H	H	H	H	H	H	H	H	H	L
L	X	X	X	X	X	X	X	L	L	L	L	L	H
L	X	X	X	X	X	X	L	H	L	L	H	L	H
L	X	X	X	X	X	L	H	H	L	H	L	L	H
L	X	X	X	X	L	H	H	H	L	H	H	L	H
L	X	X	X	L	H	H	H	H	H	L	L	L	H
L	X	X	L	H	H	H	H	H	H	L	H	L	H
L	X	L	H	H	H	H	H·	H	H	H	L	L	H
L	L	H	H	H	H	H	H	H	H	H	H	L	H

H = high logic level, L = low logic level, X = irrelevant

TEXAS INSTRUMENTS

3-323

Fig. 3.23 (a) 8 line to 3 line priority encoder (courtesy of Texas Instruments).

TYPES SN74148 (TIM9907), SN74LS148
SN54148, SN54LS148
8-LINE TO 3-LINE PRIORITY ENCODERS

logic diagram

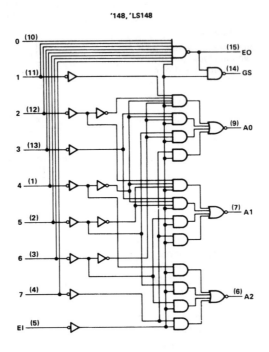

'148, 'LS148

Pin numbers shown on logic notation are for D, J or N packages.

Texas
INSTRUMENTS

Fig. 3.23 (b)   8 line to 3 line priority encoder (courtesy of Texas Instruments).

TYPES SN74148 (TIM9907), SN74LS148
SN54148, SN54LS148
8-LINE TO 3-LINE PRIORITY ENCODERS

schematics of inputs and outputs

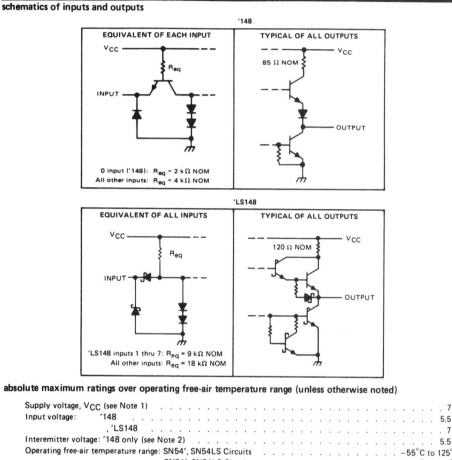

'148

EQUIVALENT OF EACH INPUT

TYPICAL OF ALL OUTPUTS

0 input ('148): R_{eq} = 2 kΩ NOM
All other inputs: R_{eq} = 4 kΩ NOM

85 Ω NOM

'LS148

EQUIVALENT OF ALL INPUTS

TYPICAL OF ALL OUTPUTS

120 Ω NOM

'LS148 inputs 1 thru 7: R_{eq} = 9 kΩ NOM
All other inputs: R_{eq} = 18 kΩ NOM

absolute maximum ratings over operating free-air temperature range (unless otherwise noted)

Supply voltage, V_{CC} (see Note 1) . . . . 7 V
Input voltage: '148 . . . . 5.5 V
              , 'LS148 . . . . 7 V
Interemitter voltage: '148 only (see Note 2) . . . . 5.5 V
Operating free-air temperature range: SN54', SN54LS Circuits . . . . −55°C to 125°C
                             SN74', SN74LS Circuits . . . . 0°C to 70°C
Storage temperature range . . . . −65°C to 150°C

NOTES: 1. Voltage values, except interemitter voltage, are with respect to network ground terminal.
        2. This is the voltage between two emitters of a multiple-emitter transistor. For '148 circuits, this rating applies between any two of
          the eight data lines, 0 through 7.

recommended operating conditions

	SN54'			SN74'			SN54LS'			SN74LS'			UNIT
	MIN	NOM	MAX	MIN	NOM	MAX	MIN	NOM	MAX	MIN	NOM	MAX	
Supply voltage, V_{CC}	4.5	5	5.5	4.75	5	5.25	4.5	5	5.5	4.75	5	5.25	V
High-level output current, I_{OH}			−800			−800			−400			400	µA
Low-level output current, I_{OL}			16			16			4			8	mA
Operating free-air temperature, T_A	−55		125	0		70	−55		125	0		70	C

TEXAS
INSTRUMENTS

3-325

Fig. 3.23 (c)   8 line to 3 line priority encoder (courtesy of Texas Instruments).

TYPES SN74LS148, SN54LS148
8-LINE TO 3-LINE PRIORITY ENCODERS

electrical characteristics over recommended operating free-air temperature range (unless otherwise noted)

PARAMETER		TEST CONDITIONS†		SN54LS' MIN	TYP‡	MAX	SN74LS' MIN	TYP‡	MAX	UNIT
V_{IH} High-level input voltage				2			2			V
V_{IL} Low-level input voltage						0.7			0.8	V
V_{IK} Input clamp voltage		V_{CC} = MIN, I_I = −18 mA				−1.5			−1.5	V
V_{OH} High-level output voltage		V_{CC} = MIN, V_{IH} = 2 V, V_{IL} = 0.8 V, I_{OH} = −400 μA		2.5	3.4		2.7	3.4		V
V_{OL} Low-level output voltage		V_{CC} = MIN, V_{IH} = 2 V, V_{IL} = V_{IL}max	I_{OL} = 4 mA		0.25	0.4		0.25	0.4	V
			I_{OL} = 8 mA					0.35	0.5	
I_I Input current at maximum input voltage	'LS148 inputs 1 thru 7	V_{CC} = MAX, V_I = 7 V				0.2			0.2	mA
	All other inputs					0.1			0.1	
I_{IH} High-level input current	'LS148 inputs 1 thru 7	V_{CC} = MAX, V_I = 2.7 V				40			40	μA
	All other inputs					20			20	
I_{IL} Low-level input current	'LS148 inputs 1 thru 7	V_{CC} = MAX, V_I = 0.4 V				−0.8			−0.8	mA
	All other inputs					−0.4			−0.4	
I_{OS} Short-circuit output current§		V_{CC} = MAX		−20		−100	−20		−100	mA
I_{CC} Supply current		V_{CC} = MAX, See Note 5	Condition 1		12	20		12	20	mA
			Condition 2		10	17		10	17	mA

NOTE 5: I_{CC} (condition 1) ist measured with inputs 7 and EI grounded, other inputs and outputs open, I_{CC} (condition 2) is
measured with all inputs and outputs open.

†For conditions shown as MIN or MAX, use the appropriate value specified under recommended operating conditions.
‡All typical values are at V_{CC} = 5 V, T_A = 25″C.
§Not more than one output should be shorted at a time.

SN54LS148, SN74LS148 switching characteristics, V_{CC} = 5 V, T_A = 25°C

PARAMETER¶	FROM (INPUT)	TO (OUTPUT)	WAVEFORM	TEST CONDITIONS	MIN	TYP	MAX	UNIT
t_{PLH}	1 thru 7	A0, A1, or A2	In-phase output			14	18	ns
t_{PHL}						15	25	
t_{PLH}	1 thru 7	A0, A1, or A2	Out-of-phase output			20	36	ns
t_{PHL}						16	29	
t_{PLH}	0 thru 7	EO	Out-of-phase output			7	18	ns
t_{PHL}				C_L = 15 pF, R_L = 2 kΩ, See Note 4		25	40	
t_{PLH}	0 thru 7	GS	In-phase output			35	55	ns
t_{PHL}						9	21	
t_{PLH}	EI	A0, A1, or A2	In-phase output			16	25	ns
t_{PHL}						12	25	
t_{PLH}	EI	GS	In-phase output			12	17	ns
t_{PHL}						14	36	
t_{PLH}	EI	EO	In-phase output			12	21	ns
t_{PHL}						23	35	

¶t_{PLH} = propagation delay time, low-to-high-level output
t_{PHL} = propagation delay time, high-to-low-level output
NOTE 4: See General Information Section for load circuits and voltage waveforms.

TEXAS INSTRUMENTS

3-327

Fig. 3.23 (d) 8 line to 3 line priority encoder (courtesy of Texas Instruments).

The '148 priority encoder has eight inputs (numbered 0 to 7), each of which is active low. Input 7 is the highest priority, while input 0 is the lowest. For normal operation, the device must be enabled by connecting the *enable input* (*EI*) input to 0 volts.

If any one of the inputs is connected to logic 0, a unique 3 bit code is produced on outputs A2, A1 and A0, allowing the active input to be easily identified. At the same time, the GS output is pulled low, which may be used to generate an interrupt request.

The device is also capable of handling simultaneous interrupt requests on two or more inputs. In this case the GS output goes to logic 0 (thus generating an interrupt request), while the outputs produce the binary code associated with the highest priority active input. Once this device has been serviced (and its input returns to logic 1), the '148 produces the binary code of the next highest active input. The GS output remains low, causing another interrupt when maskable interrupts are re-enabled. Once all external devices have been serviced, the GS output returns to logic 1.

When using a priority encoder to prioritise external interrupts, the ideal solution would be to use the 3 bit number generated, to form part of the ISR address. In this case, each 3 bit code would refer to a unique ISR, allowing a much faster response to an interrupt than the polled approach seen earlier. Figure 3.24 illustrates the prioritisation of vectored interrupts, based on *mode 2* operation with the Z80 microprocessor.

Recall that in mode 2 operation, the interrupt vector table address is formed by combining the content of the interrupt register with a byte supplied by the interrupting device. This byte is placed on the data bus in response to an interrupt acknowledge cycle, in which $\overline{M}_1$ and $\overline{IORQ}$ are simultaneously active.

By examining the circuit, it can be seen that the priority encoder supplies 3 bits of the externally supplied byte, while the remaining input lines are tied to 0 volts. The GS output is used to generate the interrupt request, while the buffer outputs are enabled by the Z80 interrupt acknowledge cycle. Listing 3.8 gives the outline of a possible program, based on this circuit.

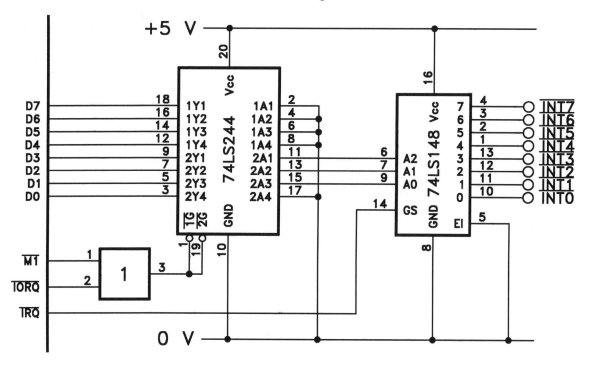

Fig 3.24 Z80 mode 2 vectored interrupts using a priority encoder.

```
ORG 4000H    ;Main program
LD SP,5000H  ;Stack = top of RAM
EI           ;Enable interrupts
IM 2         ;Mode 2
LD A,44H     ;MSB of ISR vector
LD I,A       ;table
             ;Put device init-
             ;ialisations here,
             ;+ the main program
HALT

ORG 4400H    ;ISR vector table
DEFW 4800H   ;ISR7 pointer
DEFW 4820H   ;ISR6 pointer
DEFW 4840H   ;ISR5 pointer
DEFW 4860H   ;ISR4 pointer
DEFW 4880H   ;ISR3 pointer
DEFW 48A0H   ;ISR2 pointer
DEFW 48C0H   ;ISR1 pointer
DEFW 48E0H   ;ISR0 pointer


ORG 4800H    ;ISR7 entry point
ORG 4820H    ;ISR6 entry point
ORG 4840H    ;ISR5 entry point
ORG 4860H    ;ISR4 entry point
ORG 4880H    ;ISR3 entry point
ORG 48A0H    ;ISR2 entry point
ORG 48C0H    ;ISR1 entry point
ORG 48E0H    ;ISR0 entry point
```

Listing 3.8 Structure of a Z80 mode 2 program using vectored interrupts.

It is also possible to use a priority encoder to prioritise interrupts with the 6502 microprocessor. Since the 6502 does not support vectored interrupts directly, it is necessary to use a rather indirect approach – the *indirect jump* instruction to be precise!

In this case the '148 is connected directly to an input port, as shown in figure 3.25. As before, the GS output is used to generate a maskable interrupt, while the 3 bit 'address' of the highest priority interrupt is obtained by reading the port. The byte obtained is then used to form the least significant half of the jump instruction's vector.

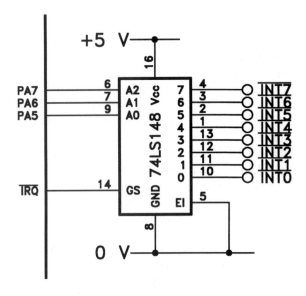

Fig. 3.25   6502 vectored interrupts using a priority encoder.

The jump indirect instruction itself would probably be placed in ROM, while the operand would point to an area of RAM. The *effective address* used by the jump instruction may then easily be altered during program execution by changing the RAM based pointer. (Details of indirect addressing may be found in *Microelectronic Systems – Book One*). Listing 3.9 shows the use of indirect addressing to implement vectored interrupts using the 6502 microprocessor.

```
        ORG $FFFE    ;Set IRQ vector
        DEFB $00,$48 ;to $4800

        ORG $4000    ;Main program
DRA:    EQU $9000    ;6502 port
DDRA:   EQU $9001    ;addresses
IVEC:   EQU $0000    ;Pointer to ISRs
        LDX #$FF      ;Initialise
        TXS           ;stack pointer
        LDA #0        ;All bits input
        STA DDRA      ;on ports A
        CLI           ;Enable interrupts
        LDA #$80      ;MSB of pointer
        STA IVEC+1
                      ;Put main program
        BRK           ;here
```

```
ORG $4800    ;IRQ ISR handler
             ;Push registers
LDA DRA      ;Read IRQ number
AND #$E0     ;Mask unwanted bits
STA IVEC     ;LSB of pointer
JMP (IVEC)   ;Go to active IRQ

ORG $8000    ;ISR7 entry point
             ;ISR7 function here
             ;Pull registers
  RTI        ;Return from IRQ

ORG $8020    ;ISR6 entry point

ORG $8040    ;ISR5 entry point

ORG $8060    ;ISR4 entry point

ORG $8080    ;ISR3 entry point

ORG $80A0    ;ISR2 entry point

ORG $80C0    ;ISR1 entry point

ORG $80E0    ;ISR0 entry point
```

Listing 3.9 6502 vectored interrupts.

The use of a priority encoder with vectored interrupts allows the microprocessor to respond quickly to an interrupt request which may have originated from one of several possible sources.

One disadvantage of this approach arises from the automatic disabling of interrupts while an interrupt is being serviced (during the ISR itself). This means that if a high priority interrupt occurs while a low priority interrupt is being serviced, this second interrupt will be ignored until the first (lower priority) ISR has been completed.

It would be useful to be able temporarily to suspend the operation of a low priority ISR, in the event of a higher priority interrupt becoming active, thus ensuring that important interrupts always receive prompt attention. The *daisy chain* interrupt system considered next (which is specific to the Z80), offers this advantage over the previously considered methods.

Daisy chain interrupts are supported by a wide range of Z80 peripheral ICs, including the *Z80 PIO (parallel input/output)*, *Z80 CTC (counter timer circuit)* and *Z80 SIO (serial input output)* components. General concepts of daisy chain interrupts are discussed here, while examples of specific devices will be left to the next chapter.

Practical Exercise	**Interrupt Priority (Priority Encoder)**

Introduction

This exercise demonstrates the connection of up to eight external (active low) interrupt signals to the microprocessor's maskable interrupt pin.

A priority encoder IC is used to prioritise the external interrupt signals and generate an interrupt request when one or more inputs is active.

Method

Connect the appropriate priority encoder circuit to your microcomputer. (Use the circuit of figure 3.24 for the Z80 or figure 3.25 for the 6502.)

Write a program and interrupt handler that will print a single letter on the screen, each time an interrupt is generated. Each input should cause a different letter to be printed.

Ensure that the program always prints the letter associated with the highest priority active interrupt.

Z80 peripheral ICs each have an interrupt request output ($\overline{\text{INT}}$), which is open drain. As seen previously, this allows several interrupt requests to be linked together, producing a wired-OR arrangement with the addition of a single pull-up resistor.

These devices are designed to operate in interrupt mode two, where the peripheral outputs a data byte in response to an interrupt acknowledge cycle. This is combined with the content of the I register to form a table address (which is an indirect pointer to the ISR). Each IC is programmed with a different 8 bit vector, ensuring that the correct ISR is executed in each case.

A potential problem with this arrangement is that two or more devices might both require service simultaneously. Bus contention would occur if both devices tried to control the data bus. For this reason, each device is provided with two additional pins, which are used to form an interrupt priority system, as shown in figure 3.26.

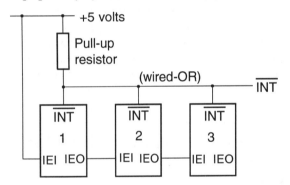

Fig. 3.26 Daisy chain interrupt priority system.

Each IC has an *interrupt enable input* (*IEI*) and an *interrupt enable output* (*IEO*). The first device has its IEI input connected to +5 volts, while its IEO output is linked to the IEI input of the next device in the chain. This arrangement is repeated until the last device is reached, whose IEO pin is left unconnected.

If a particular IC's IEI input is at logic 1, then that device is permitted (enabled) to generate an interrupt request, while a logic 0 on its IEI input inhibits the generation of interrupts. As can be seen, device 1 is permanently enabled, due to the direct connection of its IEI input to +5 volts.

In the absence of interrupt requests, the IEI and IEO pins of each IC are at logic 1, meaning that any device can generate an interrupt request. When an interrupt request occurs, the interrupting device pulls its IEO output to logic 0, which has the effect of disabling the next device in the chain. This device in turn disables its IEO output and this effect 'ripples through' the remaining devices in the chain. Figure 3.27 illustrates the logic levels on each IEI and IEO pin caused by device 1 generating an interrupt request.

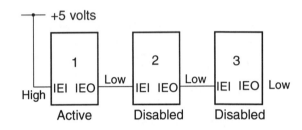

Fig. 3.27   IEI and IEO levels caused by an active interrupt request (device 1).

Once device 1 has been serviced, its IEO output returns to logic 1, and devices 2 and 3 are re-enabled.

It can be seen from this discussion that device 1 has the highest priority (since its IEI input is tied directly to +5 volts), while device 3 has the lowest priority. The IEI pin of a particular IC will only be enabled if all devices higher in the chain are enabled and inactive.

When an interrupt acknowledge cycle is detected, only a single device in the daisy chain will be both active and enabled. Hence it is this device which places its interrupt vector byte onto the data bus. Any lower priority devices are forced to wait for completion of higher priority ISR(s), at which point their IEI input goes high. Bus contention is avoided by this simple technique.

The daisy chain system also allows a partially completed ISR to be suspended (and later resumed), in the event of a higher priority interrupt becoming active during servicing of lower priority devices. This allows the microprocessor to deal with *nested* interrupt requests.

Figure 3.28 shows the timing of a Z80 interrupt acknowledge cycle.

Notice firstly that $\overline{M_1}$ and $\overline{IORQ}$ are simultaneously at logic 0 during the latter part of the cycle, which is a characteristic of an interrupt acknowledge cycle. (Compare this with an op. code fetch where $\overline{M_1}$ and $\overline{MREQ}$ are active).

The second important feature is the fact that the interrupt request line is active only until an interrupt acknowledge is detected, after which the $\overline{INT}$ input returns to its previous state. This eliminates the possibility of a single interrupt request causing multiple interrupts, due to the level sensitive nature of the $\overline{INT}$ pin.

An important conclusion that may be drawn here, is that there is no need to delay the re-enabling of maskable interrupts until the end of the ISR. Instead, if interrupts are re-enabled at the earliest possible point during the ISR, this would allow the possibility of *nested interrupts*. In other words, a low priority ISR could be temporarily suspended, if a higher priority interrupt occurred before its completion.

A final problem with nested interrupts is that the active device must recognise when its ISR has been completed successfully. The serviced device must reset its IEO pin back to logic 1, to re-enable other devices in the chain. To achieve this, the active device monitors the data bus until it detects the return from interrupt instruction (RETI), after which it relinquishes control. Devices lower in the daisy chain may then resume service, or generate an interrupt request.

Three different methods of prioritising interrupt requests have been considered in this section, each having their own advantages and disadvantages. These techniques are suitable for use with microcomputers in which the various components are physically separate. Another approach is needed in highly integrated microcomputer systems, where many functions are contained in a single integrated circuit (often called *single chip microcomputers* or *microcontrollers*).

A typical microcontroller may contain a microprocessor, ROM, RAM, PIO, CTC and SIO components, all housed within a single package. Clearly there is no possibility of determining interrupt priorities with external hardware. Instead, such systems are often provided with one or more special purpose registers, which are used to control the generation of interrupts in the system.

Each sub-section of the microcontroller normally has its own *interrupt enable register*, which is used selectively to enable or disable the generation of interrupts from that source. An *interrupt priority register* is then used to determine the relative importance of each interrupt type. Thus the designer can select the types of interrupt which are permitted, as well as their relative importance, simply by altering the software initialisation of the system.

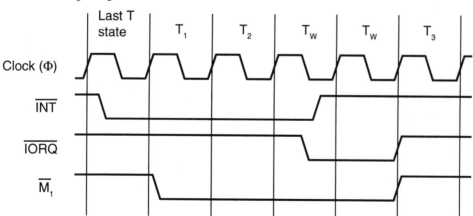

Fig. 3.28 Z80 interrupt acknowledge cycle.

Summary

The main points covered in this chapter are:

- The use of *subroutines* allows programs to be developed in a modular fashion.
- Subroutine operation is based on the *stack*, which is an area of memory used for the temporary storage of data. The stack is used to hold the subroutine's *return address*.
- Storage and retrieval of data from the stack is controlled by the *stack pointer* register, which always refers to the *top of the stack*. The stack operates as a *last-in, first-out store*.
- Subroutine calls may be *nested*, provided that the stack's capacity is not exceeded.
- *Push* and *pop* (or *pull*) instructions may be used to save temporarily the state of microprocessor data registers. This allows subroutines to be developed that leave main program registers unaltered.
- The Z80 and 6502 are provided with *maskable* and *non-maskable interrupt* inputs. These allow normal operation to be suspended in response to an external event.
- An *interrupt service routine*, which is called in response to an interrupt, is similar in structure to a subroutine.
- Registers used by the main program must not be altered by the ISR. This is achieved using *push* and *pop* (or *pull*) instructions, and allows the main program to resume as though it had not been suspended.
- The alternative to interrupts is to regularly test or *poll* the status of external devices. This is software intensive and slow, compared with an interrupt driven system.
- Systems that are connected to multiple sources of interrupt must *prioritise* these signals, so that important events always receive prompt attention.
- Interrupts may be prioritised in software (by the testing of status flags), or in hardware using a *priority encoder* or a *daisy chain* interrupt priority system.
- *Single chip microcomputers* (*microcontrollers*) may allow interrupt priority to be determined using special purpose registers.

Problems

1   Define the following terms.

   a) Subroutine
   b) Nested subroutine
   c) Stack
   d) Stack pointer

2   Explain the sequence of events that occurs when a subroutine is called from the main program.

3   Explain how the content of one or more data registers may be saved and later recovered from the stack. Why is this facility useful?

4   Define the following terms.

   a) Interrupt
   b) Nested interrupt
   c) Maskable interrupt
   c) Non maskable interrupt
   d) Vectored interrupt

5   Describe the sequence of events that occurs when a non-maskable interrupt is generated by an external event.

6   Compare the use of polling and interrupts for dealing with external devices that may require attention from time to time.

7   Explain why it is sometimes necessary to prioritise interrupts.

8   Name and describe briefly two different methods of prioritising interrupt requests.

4 Microprocessor Support ICs

Aims

When you have completed this chapter you should be able to:

1 Write programs and use hardware to transfer data to and from a microcomputer in parallel and serial forms.
2 Use commercially available serial and parallel interfaces and appreciate their typical applications.
3 Recognise the advantages of using interrupt driven software in data communications.
4 Use software and hardware techniques to measure time intervals, generate time delays and determine the current time and date.

Parallel Interface ICs

Typical PIO ICs were introduced in the first volume. Applications included the configuration of the

- entire port used for data input,
- entire port used for data output,
- bi-directional data transfer with some bits configured as inputs and others as outputs.

Although the existence of *handshake lines* to control the flow of information across a parallel interface was also mentioned in book one, actual examples of their use were not considered at that time. This section considers the use of handshaking in parallel data transfer, as well as the use of interrupts to indicate when a byte of data has been either transmitted or received.

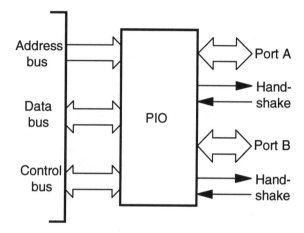

Fig. 4.1 Typical parallel input/output IC.

As shown in figure 4.1 above, each port has two handshake lines, one of which is normally an output, while the other is an input. To understand the function of these additional hanshake signals, consider the parallel connection of a computer and a printer, via a (simplified) *Centronics* interface, as shown in figure 4.2.

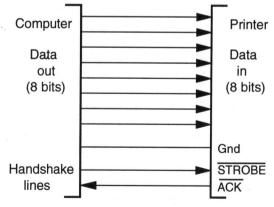

Fig. 4.2 A typical parallel printer interface.

A parallel interface is used typically to connect the computer to devices such as printers or other output devices. This type of interface is ideal in situations where the distance between the transmitter and receiver is relatively short, due to high cabling costs and the problem of crosstalk between adjacent conductors. Data transfer is normally an output from the computer, with additional lines used to indicate device status. (Bi-directional data transfer, particularly over longer distances, is usually based on serial data transfer techniques, as will be seen later.)

Peripheral devices with a Centronics interface are normally fitted with a 36 pin connector, although not all of these pins are in common use. Table 4.1 lists some of the more frequently used pin names and functions.

The BUSY line is an output from the printer, and hence an input to the computer. When this line is low it indicates the printer is ready to accept a new byte of data from the computer. When high, it inhibits the transfer of further data. Certain conditions such as *paper out* or *buffer full* may cause the printer to become 'busy' for extended periods. Some simple Centronics interfaces do not use the BUSY line, preferring to use the $\overline{\text{ACK}}$ (acknowledge) signal alone to indicate device status.

The $\overline{\text{STROBE}}$ line is an output from the computer, which is pulsed low to indicate (to the printer) that a new byte of data is available on the data lines. Once the printer has processed the incoming data, the printer indicates, by pulling the $\overline{\text{ACK}}$ signal low, that it is able to accept further data. This sequence of events is illustrated by figure 4.3.

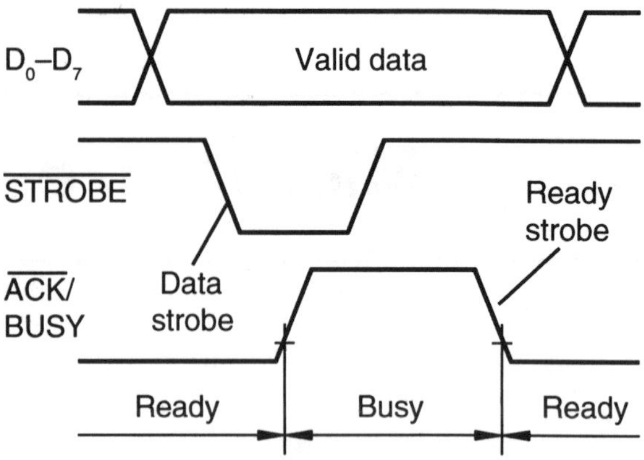

Fig. 4.3 Typical handshake signal waveforms.

In most cases the falling edge of the $\overline{\text{STROBE}}$ signal generates an interrupt to the microprocessor which is inside the printer. Similarly, the 1 to 0 transition on the $\overline{\text{ACK}}$ line may prompt the computer to send more data via an interrupt request. The use of interrupt driven communication software allows each device to concentrate on other tasks at times when the interface is inactive, which is more efficient than a polled approach.

The use of a hardware buffer may also increase the efficiency of the interface, as it allows rapid file transfer to the peripheral, which may then process the incoming data at its own pace. Printers, plotters and other mechanical devices are commonly fitted with buffers for this reason. In this case, it is normally the BUSY signal which indicates a buffer full condition, while $\overline{\text{ACK}}$ is used to control the flow of individual bytes across the interface.

Pin	Signal Name	Signal Direction	Description
1	$\overline{\text{STROBE}}$	Input to printer	A 1 to 0 transition indicates valid data.
2–9	Data lines	Inputs to printer	Eight data lines.
10	$\overline{\text{ACK}}$	Output from printer	A 1 to 0 transition acknowledges data reception.
11	BUSY	Output from printer	A high level inhibits the transfer of further data.
12	PE	Output from printer	This indicates the printer is out of paper.
19–30	GND	Supply	These signals prevent cross-talk and provide a voltage reference at each end of the interface.

Table 4.1 Commonly used Centronics printer interface pins.

Handshaking with the Z80 PIO

The Z80 PIO is addressed as a contiguous group of four input/output port locations, as shown in table 4.2.

Label	Full name	Address
DRA	Data register A	BASE
DRB	Data register B	BASE + 1
CRA	Control register A	BASE + 2
CRB	Control register B	BASE + 3

Table 4.2 Z80 PIO port addresses.

Although the base address for the PIO depends on the address decoding circuitry of the computer being used, assembler directives may be used to define any labels used. A typical example is shown in listing 4.1.

```
                        ;Initialisation
BASE:   EQU   0         ;Alter as req'd
DRA:    EQU   BASE
DRB:    EQU   BASE+1
CRA:    EQU   BASE+2
CRB:    EQU   BASE+3

                        ;Put the main
                        ;program here
```

Listing 4.1 General port initialisation routine.

In this case, altering the first line causes the address of each PIO register to be recalculated automatically during assembly.

Before moving on to examine the use of handshake lines and the generation of interrupts, it is useful to recap briefly the PIO's internal operation.

Practical Exercise   Centronics Printer Interface

Introduction

The purpose of this exercise is to study an industry standard parallel printer interface. Emphasis is placed on understanding the use of handshake signals to convey timing and status information.

Method

a) Connect the computer and printer via the parallel cable.

b) Based on the computer or printer manual, identify the pins of interest and (carefully) connect them to the test instrument.

c) Record typical data transfer waveforms, obtained while printing a test document.

Equipment required

Microcomputer with a Centronics interface.
Printer and parallel printer cable.
Dual channel storage oscilloscope.
Logic analyser.

Results

a) Determine the two voltage levels used by the interface to represent logic 1 and 0.

b) Draw a diagram, showing the timing of the handshake lines ($\overline{\text{STROBE}}$, $\overline{\text{ACK}}$ and BUSY), indicating the significance of each section.

c) If you have a logic analyser, compare the data on the data pins with the ASCII codes of the characters printed.

The Z80 PIO can operate in one of four possible modes, as shown in table 4.3.

Mode	Function
0	Byte output mode
1	Byte input mode
2	Byte input/output mode
3	Bit input/output mode

Table 4.3 Z80 PIO operating modes.

An operating mode is selected by writing a *mode control word* to the control register of the appropriate port (A or B). The structure of this control word is shown by figure 4.4.

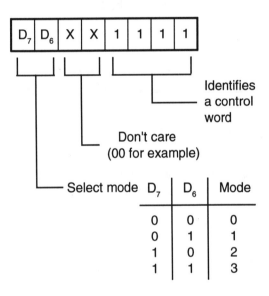

Fig. 4.4 Mode control word internal structure.

Thus a mode control word is identified by the presence of 1111_2 in the least significant half of the byte which is stored in the control register.

The same control register may also be used for other functions, such as enabling or disabling of interrupts or specifying an interrupt vector. In each case, a particular bit pattern is used to identify the control function. (In some cases, the meaning of information stored in the control register may depend on the order in which it is written, as well as the actual binary content.)

Each port has two associated handshake lines which are used to control the flow of information, and which may optionally generate an interrupt when data is transmitted or received. The names and functions of each handshake line are given in table 4.4.

Pin	Direction	Function
ARDY	Output	Port A ready
ASTB	Input	Port A strobe
BRDY	Output	Port B ready
BSTB	Input	Port B strobe

Table 4.4 Z80 PIO handshake lines.

The exact function of each handshake line depends on the mode of operation.

In byte output mode (mode 0), the READY signal goes high to indicate that valid data is available on the port. The peripheral device indicates it has received the data with a low to high transition on the $\overline{\text{STROBE}}$ line, as shown in figure 4.5.

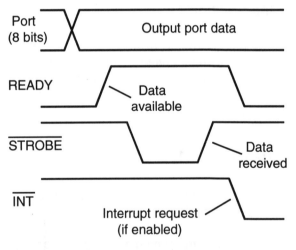

Fig. 4.5 Output mode signal timing.

The Z80 PIO may optionally generate an interrupt request, in response to the external device acknowledging data reception. This interrupt prompts the microprocessor to transmit further data, if available.

Byte input mode (mode 1), may also use hand-shake signals to control the flow of information and optionally generate interrupt requests. In this case a 0 to 1 transition of the $\overline{\text{STROBE}}$ signal is used by the external device to latch data into the Z80 PIO's data register. If enabled, this causes the PIO to generate an interrupt request.

At the same time, the READY signal goes low to indicate that the input data register is full. Once the Z80 has performed a read of the data register, the READY signal goes high, indicating that the PIO is once again ready to receive data. This process is illustrated by figure 4.6.

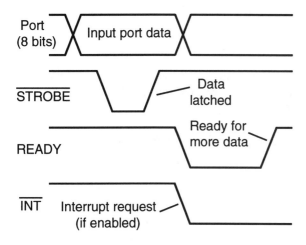

Fig. 4.6 Input mode signal timing.

Assuming that interrupts are to be used, several initialisation steps are required. First an *interrupt enable word* must be written into the control register of the appropriate port. The format of this interrupt enable word is shown in figure 4.7.

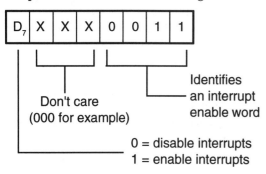

Fig. 4.7 Interrupt enable word.

As was seen in the last chapter, the Z80 can operate in one of several different interrupt modes. The Z80 PIO is designed to operate with interrupt mode 2 and is therefore capable of generating an interrupt vector byte in response to a Z80 interrupt acknowledge cycle. This allows each interrupt source to have its own interrupt service routine, the address of which is specified by the peripheral. (Recall also that Z80 peripheral ICs also use the daisy chain interrupt priority system.)

For correct interrupt mode 2 operation, four initialisation steps must be performed.

- enable interrupts.
- select interrupt mode 2.
- write the high byte of the interrupt vector table address into the I register.
- write the low byte of the interrupt vector table address into the appropriate PIO control register.

An *interrupt vector word* is identified by the fact that the least significant bit of the byte written into the control register is zero, as shown in figure 4.8.

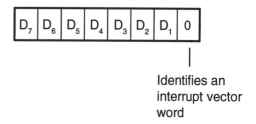

Fig. 4.8 Interrupt vector word.

Example

Write a program that will output a zero-terminated ASCII string of characters from port A of a Z80 PIO The program should demonstrate the use of hanshake signals to control the flow of data across the interface.

A solution to this problem is shown in listing 4.2, which makes use of byte output mode (mode 0) and interrupt mode 2 (vectored interrupts).

```
        ORG 4000H     ;Data definition
DTA: DEFB 'Test message',0
                      ;Main program
     LD SP,5000H ;Stack = top of RAM
     EI            ;Enable interrupts
     IM 2          ;Mode 2
     LD A,44H      ;MSB of ISR vector
     LD I,A        ;table   = 44H
                   ;PIO initialisation
BSE: EQU  0        ;Alter as req'd
DRA: EQU  BSE
CRA: EQU  BSE+2
     LD A,0FH      ;Byte output mode
     OUT (CRA),A ;(mode 0)
     LD A,83H      ;Enable PIO IRQ
     OUT (CRA),A
     LD A,0        ;LSB of interrupt
     OUT (CRA),A ;vector = 00H
                   ;Put the main
                   ;program here
PRT: LD HL,DTA     ;Start of data blk
     LD A,(HL)
     OUT (DRA),A ;Send data to port
                   ;Put rest of main
                   ;program here...

     HALT

     ORG 4400H     ;ISR vector table
     DEFW 4800H    ;(Only one entry)

     ORG 4800H     ;ISR entry point
     PUSH AF       ;Push registers
     INC HL        ;Increment pointer
     LD A,(HL)     ;Load next char.
     CP 0          ;End of string (0)?
     JP Z,FIN      ;Don't send if yes
     OUT (DRA),A ;Send data to port
FIN: POP AF        ;Pop registers
     EI            ;Enable interrupts
     RETI          ;Return from INT
```

Listing 4.2 Byte output mode with interrupts.

The main program begins by defining an ASCIIZ (zero terminated) string, which is to be transmitted across the interface, after which initialisation of the microprocessor and PIO are performed.

A memory pointer (the HL register pair) is then set to the first character of the string, which is transmitted via port A. Subsequent bytes are then transmitted in response to interrupt requests generated by the receiver, which allows the main program to concentrate on other tasks.

Each time the receiving device acknowledges reception (with a low to high transition on the STROBE line), an interrupt request occurs and the ISR at address 4800_{16} is called. The ISR loads the next ASCII character and, provided this is not zero, transmits the character via port A. When the end of the string is detected, the ISR terminates (and may also disable further interrupts, if required).

Example

Write a program that uses handshake signals and interrupts to receive input data from port A of a Z80 PIO.

A solution to this problem is shown in listing 4.3 below.

```
DTA: ORG 4000H     ;Text buffer
                   ;= 40 characters
     ORG 4040H     ;Main program
     LD SP,5000H ;Stack = top of RAM
     EI            ;Enable interrupts
     IM 2          ;Mode 2
     LD A,44H      ;MSB of interrupt
     LD I,A        ;vector = 44H
                   ;PIO initialisation
BSE: EQU  0        ;Alter as req'd
DRA: EQU  BSE
CRA: EQU  BSE+2
     LD A,4FH      ;Byte input mode
     OUT (CRA),A ;(mode 1)
     LD A,83H      ;Enable PIO IRQ
     OUT (CRA),A
     LD A,0        ;LSB of interrupt
     OUT (CRA),A ;vector = 00H
     LD HL,DTA     ;Start of data blk
                   ;Put rest of main
                   ;program here...
     HALT
```

continued overleaf...

```
ORG 4400H    ;ISR vector table
DEFW 4800H    ;(Only one entry)

ORG 4800H    ;ISR entry point
PUSH AF      ;Push registers
IN A,(DRA)    ;Read incoming data
LD (HL),A    ;Put in text buffer
INC HL       ;Increment pointer
POP AF       ;Pop registers
EI           ;Enable interrupts
RETI         ;Return from INT
```

Listing 4.3 Byte input mode with interrupts

Program initialisation is similar to that of the previous program, except that mode 1 (byte input) is used, rather than mode 0 (byte output).

Each time a character is received via port A, an interrupt request is generated. This causes the ISR at address 4800_{16} to be called, which stores the incoming data in a text buffer (an area of memory set aside to hold data that has not yet been processed by the computer). After each byte has been stored, the HL register pair is incremented.

(It is assumed in these two examples that the HL register pair is not altered by the main program. If HL is required by the main program, a memory based pointer may be used instead).

This section has concentrated on modes 0 and 1 of the Z80 PIO. Those readers who require further information on the Z80 PIO, including modes 2 and 3, are referred to the *Zilog Intelligent Peripheral Controllers Manual* (1991).

Practical Exercise   Parallel Computer Interface

Method

a) Connect two computers, as shown in the diagram opposite.

b) Write software for the transmitting computer which causes data to be transmitted via port B, making use of the handshake lines and interrupts.

c) Write software for the receiving computer which allows incoming data to be received via port B. This software should make use of the handshake lines and be interrupt driven. Ideally, the software should per form some visible action, such as displaying the data on screen.

d) Use suitable test equipment to observe the flow of data across the interface.

Circuit Connections

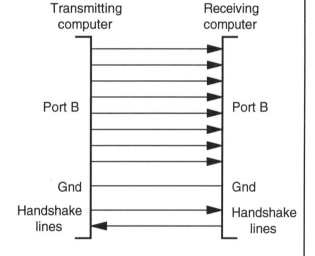

Handshake Connections

Z80 - $\overline{\text{STROBE}}$ to READY and vice versa.
6502 - CB1 to CB2 and vice versa.

Practical Exercise   Local Area Network

Introduction

This exercise examines the connection of several computers via serial or parallel interfaces. This arrangement allows the transfer of data or commands between computers and is known as a *local area network* or *LAN.*

Method

a) Connect several computers as shown, via serial or parallel interfaces (use serial connection over longer distances).

b) Each computer should be allocated a unique *node address* or *identifier.*

   (A 4 bit address will allow up to 16 computers to be connected to the LAN).

c) Devise a data format where each byte (or *data packet*) transmitted contains a mixture of address (4 bits) and data (4 bits).

d) Write interrupt driven software to transmit and receive data via each node. Received data that is not for the current node should be retransmitted (unaltered).

Conclusions

Discuss the usefulness of such a LAN for

- Remote computer control
- Peripheral sharing (such as printers)
- Remote data storage
- Parallel processing (sharing complex tasks between computers)

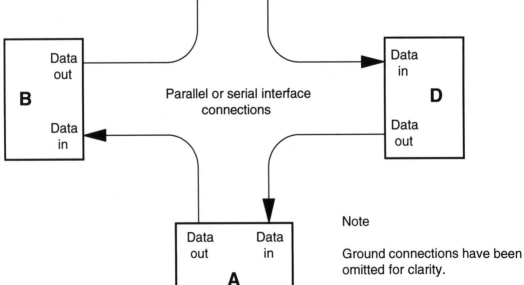

Parallel or serial interface connections

Note

Ground connections have been omitted for clarity.

Handshaking with the 6522 VIA

The 6522 *versatile interface adaptor* (*VIA*) is a commonly used 6502 peripheral which contains

- two 8 bit input/output ports with handshake lines,
- two 16 bit timer/counters,
- an 8 bit shift register.

The timer/counter and shift register features of this IC will be examined shortly, while this section concentrates on parallel data transfer with handshaking. Figure 4.9 shows the external connections to the 6522.

As can be seen, each 8 bit port has a pair of handshake lines (CA1 and CA2 for port A and CB1 and CB2 for port B). CA1 and CB1 always act as inputs, while CA2 and CB2 may act either as outputs or inputs. In fact, the 6522 handshake lines are extremely flexible and may be configured to operate in wide variety of modes. For example, both input and output handshake lines may be programmed to be either active high or active low, greatly simplifying any interfacing task.

Four address lines are used to select one of the 16 internal registers. These registers appear as a contiguous group of memory locations, as selected by the address decoding circuitry.

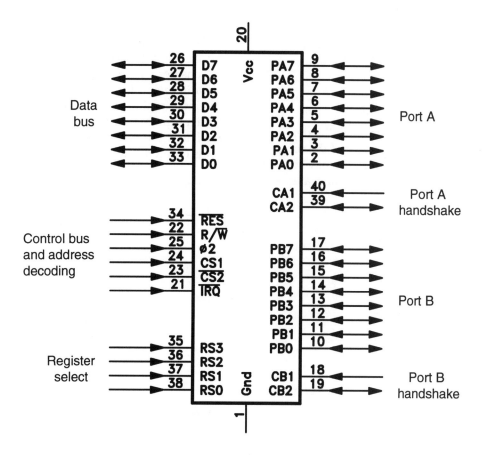

Fig. 4.9 6522 schematic layout.

The names and basic functions of each register are given in table 4.5.

Offset	Register	Description
0	DRB	Data register B
1	DRA	Data register A
2	DDRB	Data direction register B
3	DDRA	Data direction register A
4	TIC-L	Timer 1 counter/latch (low)
5	TIC-H	Timer 1 counter (high)
6	TIL-L	Timer 1 latch (low)
7	TIL-H	Timer 1 latch (high)
8	T2C-L	Timer 2 counter/latch (low)
9	T2C-H	Timer 2 counter (high)
10	SR	Shift register
11	ACR	Auxiliary control register
12	PCR	Peripheral control register
13	IFR	Interrupt flag register
14	IER	Interrupt enable register
15	DRA-N	As DRA but without handshaking

Table 4.5 6522 internal registers.

Although the base address for the VIA depends on the address decoding circuitry of the computer being used, assembler directives may be used to define any labels used. A typical example is shown in listing 4.4

```
                ;Initialisation
BASE:   EQU $8000   ;Alter as req'd
DRB:    EQU BASE
DRA:    EQU BASE+1
DDRB:   EQU BASE+2
DDRA:   EQU BASE+3
ACR:    EQU BASE+11
PCR:    EQU BASE+12
IFR:    EQU BASE+13
IER:    EQU BASE+14
                ;Put the main
                ;program here
```

Listing 4.4 General port initialisation routine.

In this case, altering the first line causes the address of each VIA register to be recalculated automatically during assembly.

Before moving on to examine the use of handshake lines and the generation of interrupts, it is useful to briefly recap the methods used to input and output data via ports A and B.

Each port has a *data register* and a *data direction register*. Each bit in the data direction register defines whether the same bit in the data register acts as an input or an output. Writing a logic 1 to a particular bit configures that line as an output while a logic 0 configures it as an input. Thus storing $0F_{16}$ in data direction register A (DDRA) configures the upper 4 bits of port A as inputs and the lower 4 bits as outputs. Data may be output from port A by writing information to data register A (DRA), and may be read by reading DRA.

In addition to the two data registers (DRA and DRB) and their data direction registers (DDRA and DDRB), several other registers are important when using handshake lines. These include the *auxiliary control register* (ACR) and the *peripheral control register* (PCR).

Considering the ACR first, only the least significant two bits are important here. (The remainder of this port is used by the two timer/counters and the shift register, which are considered later.) Fig. 4.10 shows the section of the PCR that is of interest for handshake control.

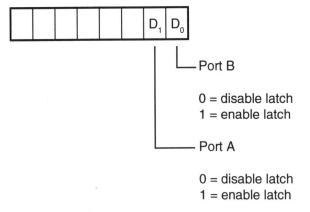

Fig. 4.10 Auxiliary control register input latch control.

When latching is enabled, an active transition on the CA1 or CB1 line causes the data present on the port inputs to be latched into the data register.

If latching is disabled, the data input is that present on the input pins at the moment when the microprocessor reads the input port. As will be seen shortly, an active transition on any one of the handshake lines may also be used to generate an interrupt request (if enabled) or simply set a status flag.

The PCR is used to configure the detailed operation of the hanshake lines, as shown in figure 4.11.

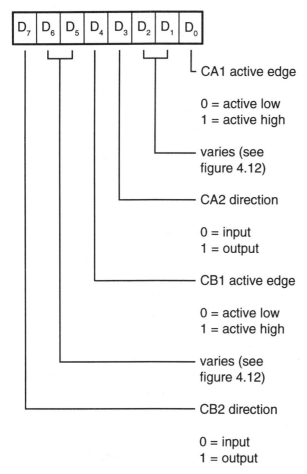

Fig. 4.11 Peripheral control register.

The upper half of the PCR is used to configure the port B hanshake lines, while the lower half controls port A. As mentioned previously, input data may be latched on an active transition of CA1 or CB1. Bits 0 and 4 of the PCR allow these active transitions to be configured as rising or falling edges.

Bits 3 and 7 are used to alter the direction of handshake lines CA2 and CB2. The exact purpose of bits 1 and 2 for port A, together with bits 5 and 6 for port B depend on the direction of CA2 and CB2 respectively. The function of these bits is shown by figure 4.12.

When CA2/CB2 is an input...

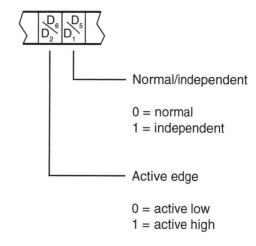

When CA2/CB2 is an output...

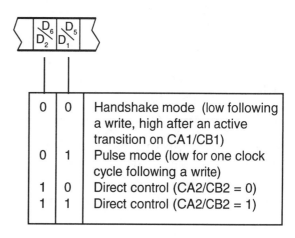

Fig. 4.12 PCR bits 1 + 2 and 5 + 6 functions.

Consider first the case where CA2 or CB2 is an input: the left-hand bit may be used to set the active edge to be active low or active high. The right-hand bit controls whether this handshake line operates in *normal* or *independent* mode.

An active transition on any handshake line causes the relevant bit in the *interrupt flag register* (*IFR*) to be set (as will be seen shortly), and may optionally cause an interrupt request to be generated. *Normal mode* indicates that a flag set by an active transition on one of the handshake lines will be cleared automatically once the associated data register has been written to or read from. This mode is generally used during handshake controlled data transfers, as the programmer does not have to clear the IFR manually following each data transfer. *Manual mode* indicates that a flag that has been set by an active handshake line transition remains set until it is cleared by the programmer.

When CA2 or CB2 is an output, these two PCR bits are used to set one of four possible output modes. *Handshake* and *pulse* modes allow the programmer to select two different output handshake waveforms, as shown in figure 4.13(a) and 4.13(b).

Pulse output waveforms...

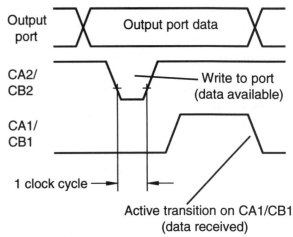

Fig. 4.13(b) Pulse mode output waveforms on CA2 or CB2.

(In the above examples it has been assumed that the active transition of CA1 or CB1 is active low, although this can be altered by the programmer, if required.)

In cases where the port is configured as an input, handshake lines CA2 or CB2 may be directly controlled to indicate the computer's readiness to receive data. For example a 1 to 0 transition on CA2 or CB2 might indicate to an external device that the previously transmitted data byte has been received. An active transition on CA1 or CB1 would then indicate the presence of new data, possibly generating an interrupt request.

The *direct control* facility which is available when CA2 or CB2 are configured as outputs, may be used to generate the required handshake signal transition. In this case the program must directly set the output handshake line to logic 1 or logic 0 at each stage, thus generating an appropriate handshake waveform and timing.

Hanshake output waveforms...

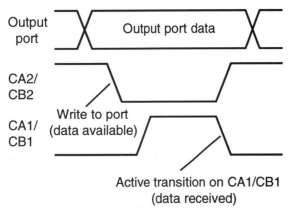

Fig. 4.13(a) Handshake mode output waveforms on CA2 or CB2.

Care should be taken when altering one or more bits of a VIA register, not to alter accidentally bits that may be used by other programs. The safest way to alter part of a register is to first load the register, then use bit masking techniques to alter the required section. Finally the modified register content may be written back. As an example listing 4.5 demonstrates 'safe' and 'unsafe' ways of setting a single bit in the PCR.

```
            ;Unsafe method
LDA #$80    ;Set MSB

STA PCR
            ;(and clear other
            ;bits)

            ;Safe method
LDA PCR     ;Get old PCR
ORA #$80    ;Set MSB
STA PCR     Save new PCR
```

Listing 4.5 Methods of modifying a VIA register.

Similarly one or more bits may be safely cleared using a logical AND instruction, as shown in listing 4.6.

```
LDA PCR     ;Get old PCR
AND #$7F    ;Clear MSB
STA PCR     ;Save new PCR
```

Listing 4.6 Clearing a single bit of a register.

Example

Demonstrate the configuration of port B as an output port with hanshake line CB1 configured as an active low input and CB2 programmed as an active low pulse (generated by a write to data register A). Port B should be unaffected.

```
LDA #$FF
STA DRA   ;Port A set as output
LDA PCR   ;Get old PCR
AND #$0F  ;Clear left half
ORA #$A0  ;1010 in left half
```

As mentioned earlier, an active transition on one of the handshake lines will set the appropriate bit in the interrupt flag register (IFR). Figure 4.14 shows those parts of the IFR that are used by the port A and B handshake lines.

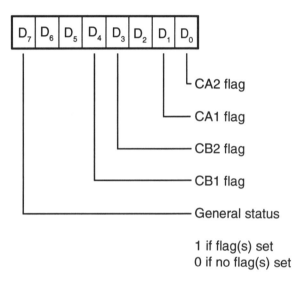

1 if flag(s) set
0 if no flag(s) set

Fig. 4.14 Interrupt flag register (IFR).

The general status bit indicates whether any flags have been set. This allows a simple software polling algorithm to be developed which begins by checking the most significant bit. Only if this is set, must further tests be performed to identify the active flag(s).

An active flag may be cleared (rather strangely) by writing a 1 into the appropriate bit of the IFR, as shown in listing 4.7.

```
LDA #$09
STA IFR    ;Clear CA2 and CB2
           ;flags directly
```

Listing 4.7 Direct clearing of IFR flags.

(This is most likely to be used in cases where CA2 or CB2 have been programmed to operate in *independent* (rather than *normal*) mode, as was explained during discussion of the PCR.)

Each bit in the IFR corresponds exactly with the same bit in the *interrupt enable register (IER)*, as shown in figure 4.15.

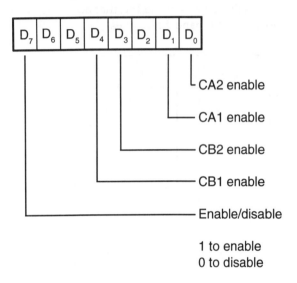

1 to enable
0 to disable

Fig. 4.15 Interrupt enable register (IER).

(Once again, bits used by other section of the VIA have been omitted here, but will be discussed later.)

A particular bit in the IER determines whether or not an interrupt request will be generated if the same bit in the IFR becomes set. Thus the content of this register determines which sources of interrupt will be enabled or disabled.

The procedure for enabling and disabling particular interrupt sources is once again, rather unusual. When a byte is written to the IER, the effect of a 1 in a particular bit position depends on the state of the most significant bit. If the MSB is 1 then writing a 1 into a bit position enables the associated interrupt source. When the MSB is 0, writing a 1 causes that interrupt source to be disabled. (Writing 0 into a bit position has no effect.) Listing 4.8 demonstrates the enabling of the CA1 interrupt and the disabling of all others.

```
LDA #$7D  ;0111 1101 in binary
STA IER   ;MSB=0 (1s disabled)
LDA #$82  ;1000 0010 in binary
STA IER   ;MSB=1 (1s enabled)
```

Listing 4.8 Enabling the CA1 interrupt.

Example

Write a program that outputs a zero-terminated string of ASCII characters from port A of a 6522 VIA. The program should demonstrate the use of handshake signals to control the flow of data across the interface.

A solution to this problem is shown in listing 4.9.

```
         ORG  $FFFE     ;Set IRQ vector
         DEFB 00,$48    ;to $4800
         ORG  $4000     ;Data definition
DTA:     DEFB 'Test message',0
BSE:     EQU  $8000     ;Port definitions
DRA:     EQU  BSE+1
DDRA:    EQU  BSE+3
PCR:     EQU  BSE+12
IFR:     EQU  BSE+13
IER:     EQU  BSE+14

                        ;Set up 6502+6522
         LDX  #$FF      ;Initialise
         TXS            ;stack pointer
         LDA  #$FF
         STA  DDRA      ;All bits output
         LDA  PCR       ;Get old PCR
         AND  #$F0      ;Clear right half
         ORA  #08       ;1000 in right half
         STA  PCR       ;Save new PCR
                        ;CA2=O/P, CA1=low
                        ;Handshake mode
         LDA  #$7F      ;0111 1111
         STA  IFR       ;Clear all flags
         LDA  #$7D      ;0111 1101
         STA  IER       ;Disable interrupts
         LDA  #$82      ;1000 0010
         STA  IER       ;Enable CA1 IRQ
         CLI            ;Enable interrupts
                        ;Main program
         LDX  #0
         LDA  DTA,X     ;Start of data blk
         STA  DRA       ;Send data to port
                        ;Put rest of main
LP:      JMP  LP        ;program here...
```

Continued overleaf...

```
        ORG  $4800    ; ISR entry point
        PHA            ; Push register(s)
        INX            ; Increment pointer
        LDA  DTA,X     ; Load next char.
        CMP  #0        ; End of string (0)?
        BEQ  FIN       ; Don't send if yes
        STA  DRA       ; Send data to port
FIN:    PLA            ; Pull register(s)
        RTI            ; Return from IRQ
```

Listing 4.9 Outputting an ASCII string (with handshaking and interrupts).

The main program begins by defining an ASCIIZ (zero terminated) string, which is to be transmitted across the interface, after which initialisation of the microprocessor and VIA are performed.

The first character of the string is then transmitted via port A. Indexed addressing is used, with the X register holding the displacement byte. Subsequent bytes are then transmitted in response to interrupt requests generated by the receiver, which allows the main program to concentrate on other tasks.

Each time the receiving device acknowledges reception (with a high to low transition on the CA1 handshake line), an interrupt request occurs and the ISR at address 4800_{16} is called. The ISR loads the next ASCII character and, provided this is not zero, transmits the character via port A. When the end of the string is detected, the ISR terminates (and may also disable further interrupts, if required).

(Notice that the X register is shared by the main program and the ISR, due to the use of indexed addressing. If this is a problem, use a spare memory location to hold the displacement byte).

Example

Write a program which uses handshake signals and interrupts to receive input data from port A of a 6522 VIA.

A solution to this problem is shown in listing 4.10.

```
        ORG  $FFFE     ; Set IRQ vector
        DEFB 00,$48    ; to $4800
        ORG  $4000     ; Text buffer
                       ; = 40H characters
        ORG  $4040
DTA:    DEFB 'Test message',0
BSE:    EQU  $8000     ; Port definitions
DRA:    EQU  BSE+1
DDRA:   EQU  BSE+3
ACR:    EQU  BSE+11
PCR:    EQU  BSE+12
IFR:    EQU  BSE+13
IER:    EQU  BSE+14
                       ; Set up 6502+6522
        LDX  #$FF      ; Initialise
        TXS            ; stack pointer
        LDA  #$00
        STA  DDRA      ; All bits input
        LDA  ACR       ; Get old ACR
        ORA  #02       ; Enable CA1 latch
        STA  ACR       ; Save new ACR
        LDA  #$7F      ; 0111 1111
        STA  IFR       ; Clear all flags
        LDA  PCR       ; Get old PCR
        AND  #$F0      ; Clear right half
        ORA  #$0D      ; 1101 in right half
        STA  PCR       ; Save new PCR
                       ; CA2=O/P, CA1=low
                       ; CA2=0 (direct con)
        LDA  #$7D      ; 0111 1101
        STA  IER       ; Disable interrupts
        LDA  #$82      ; 1000 0010
        STA  IER       ; Enable CA1 IRQ
        CLI            ; Enable interrupts
                       ; Main program
        LDX  #0
        LDA  DTA,X     ; Start of data blk
        STA  DRA       ; Send data to port
        LDA  PCR       ; Get old PCR
        ORA  #02       ; CA2=1 (direct con)
        STA  PCR       ; Save new PCR
        LDX  #0        ; Displacement=0
                       ; Put rest of main
LP:     JMP  LP        ; program here...
```

Continued overleaf...

```
       ORG $4800     ISR entry point
       PHA           ;Push register(s)
       LDA PCR       ;Get old PCR
       AND #$FD      ;CA2=0 (direct con)
       STA PCR       ;Save new PCR
       LDA DRA       ;Read incoming data
       STA DTA,X     ;Put in text buffer
       INX           ;Increment pointer
       LDA PCR       ;Get old PCR
       ORA #02       ;CA2=1 (direct con)
       STA PCR       ;Save new PCR
FIN:   PLA           ;Pull register(s)
       RTI           ;Return from IRQ
```

Listing 4.10 Inputting data with handshaking.

Program initialisation is similar to that of the previous program, except that port A is configured as an input port and the state of handshake line CA2 is directly controlled by the program. (It is assumed here that a 0 to 1 transition on CA2 indicates the computer's readiness to receive new data.)

Each time a character is received via port A (indicated by an active low transition on CA1), an interrupt request is generated. This causes the ISR at address 4800_{16} to be called, which stores the incoming data in a text buffer (an area of memory set aside to hold data which has not yet been processed by the computer). After each byte has been stored, the X register is incremented.

It is assumed in this example that the main program performs some processing task on the data buffer, which is not shown here.

Serial Data Transfer

Although parallel data transfer is frequently used for data communication over short distances, this transmission method has a number of disadvantages which prevent its use over longer cable lengths.

- High cabling costs due to the need for multiple conductors.
- Interference between signals on adjacent conductors, making communication difficult over distances greater than 2 metres. (The common use of TTL logic levels in parallel interfaces also contributes to this problem, because of its low noise immunity.)
- A Centronics interface is generally used for data transfer in one direction only.
- Data communication across the telephone network (or any other single channel), is generally in serial form.

In serial data transfer, a byte of data is transmitted across a single communication channel one bit at a time, as shown in figure 4.16.

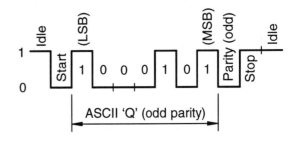

Fig. 4.16 Serial data transfer (asynchronous).

In the absence of transmitted data, the line remains in an *idle* (logic 1) state. The presence of an incoming data character is indicated by the line going to logic 0 for a brief interval. This *start bit* not only indicates that data will follow, but it supplies a *timing reference* which the receiver can use to recover the incoming data. (This timing reference applies only for the current character, as each data word is preceded by its own start bit.)

Notice here that the receiver has no advance warning of the incoming data, other than the presence of a start bit. This method of communication is known as *asynchronous*, due to the lack of any absolute timing reference. In *synchronous* communication (which is not considered in this text), the timing of incoming characters is known in advance.

After the start bit, each data bit is transmitted at regular intervals of time. In the example shown the ASCII character 'Q' is transmitted, with the least significant bit sent first. Both the transmitter and receiver must agree on the rate at which bits are to be transmitted, to allow proper reception. This is known as the *baud rate*, which is the number of bits transmitted per second. Commonly used baud rates in data communication are 300, 1200, 2400, 4800 and 9600 baud, although higher data rates are being increasingly used.

Following the 7 bit ASCII code is a *parity bit* which is used for error detection purposes. This bit, if it is enabled, may have either *odd* or *even* parity. A binary number is said to have *even parity* if it has an even number (0, 2, 4 etc.) of ones, and *odd parity* if it has an odd number (1, 3, 5 etc.) of ones. Table 4.6 gives examples of binary numbers with odd and even parity.

D_3	D_2	D_1	D_0	Parity
0	0	1	0	Odd
1	0	0	1	Even
0	1	1	1	Odd
0	0	1	1	Even

Table 4.6 Examples of odd and even parity.

During data transmission an extra bit is added to each data word in order to give either odd or even parity. Each character transmitted will then have the same type of parity.

If, during transmission, a single bit error occurs (a 1 changes to a 0 or vice versa), this will inevitably cause a change in the parity of the incoming data. This *parity error* may then cause the receiver to ask for that section of data to be retransmitted.

Immediately following the parity bit is the *stop bit*, which marks the end of the data character. This bit is normally a logic 1 and is followed by the line returning to an idle state. The channel remains idle until another character is transmitted, as indicated by another start bit.

To minimise the chance of an error during data reception, the receiver samples the incoming data at the mid-point of each bit period, as shown in figure 4.17.

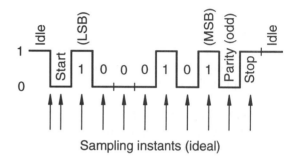

Fig. 4.17 Sampling of incoming serial data.

The first sampling instant is at the falling edge of the start bit. To eliminate the possibility of false triggering due to noise, the receiver then waits for half a bit period before checking the start bit again. If this is not at logic 0 then the receiver is reset and waits for another start bit.

Assuming that a valid start bit has been received, the remaining bits are then sampled at intervals of one bit period. This sampling technique means that small differences in frequency between the transmitter and receiver clocks can be tolerated, without generating transmission errors.

The process of dividing incoming data into groups of bits is known as *framing*. If the receiver mistakes a data bit for a start bit, this will lead to incorrect data reception, which is known as a *framing error*.

A special electronic circuit called a *shift register* is used to convert data from serial to parallel form. As each clock pulse is applied to the shift register, data moves one place to the right, while incoming serial data moves into the left-hand position. After eight clock pulses the shift register contains the complete character in parallel form.

Practical Exercise   Parity Generation and Checking

Introduction

This exercise introduces the generation and testing of parity bits for the detection of single bit data transmission errors.

Method

a) Construct the *parity generator* circuit and complete the truth table opposite.

b) Test the operation of the *parity checker* circuit by applying input data with both odd and even parity.

Parity Generator Circuit (Transmitter)

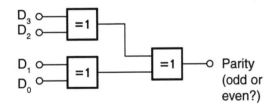

D_3
D_2 =1

D_1
D_0 =1 → =1 ○ Parity (odd or even?)

Parity Checker Circuit (Receiver)

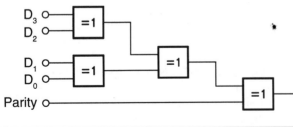

D_3
D_2 =1

D_1
D_0 =1 → =1

Parity ○ ──────────── =1 ○ Parity check

Conclusions

a) What kind of parity does the parity generator circuit produce?

b) How could the parity checker circuit be used to detect single bit data errors?

Exclusive-OR Truth Table

A	B	Q
0	0	0
0	1	1
1	0	1
1	1	0

Parity Generator Truth Table

D_3	D_2	D_1	D_0	Parity bit	Parity
0	0	0	0	0	Even
0	0	0	1		
0	0	1	0		
0	0	1	1		
0	1	0	0		
0	1	0	1		
0	1	1	0		
0	1	1	1		
1	0	0	0		
1	0	0	1		
1	0	1	0		
1	0	1	1		
1	1	0	0		
1	1	0	1		
1	1	1	0		
1	1	1	1		

Once a complete character has been received, it is normally transferred to a holding register. At the same time a flag is set in the receiver status register and an interrupt request may optionally be generated. The microprocessor is then expected to read the incoming data character before it is overwritten by the next incoming character. This arrangement is shown in figure 4.18.

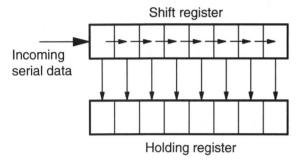

Fig. 4.18 Receiver block diagram.

The use of a holding register allows the shift register to receive new serial data, while the previous character is temporarily stored. This arrangement (known as a *double buffered* interface) gives the microprocessor a short interval in which to read the content of the holding register before its content is overwritten by new data.

A final type of error which may occur with this type of interface is a *data overrun* (or *data overflow*) error. This indicates that a character has been lost, due to the microprocessor failing to read the holding register before it is updated by incoming data.

Integrated circuits which perform asynchronous serial communication are commonly referred to as *universal asynchronous receiver/transmitters* (*UARTs*) or *asynchronous communication interface adaptors* (*ACIAs*). (ICs that also allow synchronous data transfer are called *USARTs*, which stands for *universal synchronous/asynchronous receiver/transmitter*.) UARTs and ACIAs are commonly associated with the *RS232* serial interface, which will be considered shortly.

Software Serial to Parallel Conversion

An alternative to using special purpose ICs is to use software algorithms to convert data between serial and parallel forms. A single bit of an 8 bit port may then be used to implement a serial interface in one direction, with a second bit used for communication in the opposite direction. This arrangement is shown in figure 4.19.

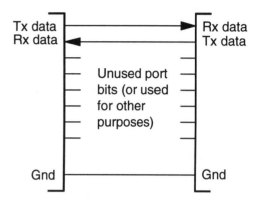

Fig. 4.19 Software serial interface.

This is easier than might be imagined, due to the inclusion of *shift* and *rotate* instructions in the microprocessor instruction set. The operation of these instructions is made possible by a general purpose shift register, which is contained within the arithmetic and logic unit of the microprocessor. (Recall that shift and rotate operations were introduced in the first volume.)

As an example, consider the operation of an instruction that shifts the content of the accumulator to the right, inserting 0 at the left hand side and placing the previous content of the LSB into the carry flag. This is shown diagrammatically in figure 4.20.

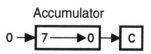

Fig. 4.20 Shift right (logical) instruction.

Practical Exercise   Shift Register

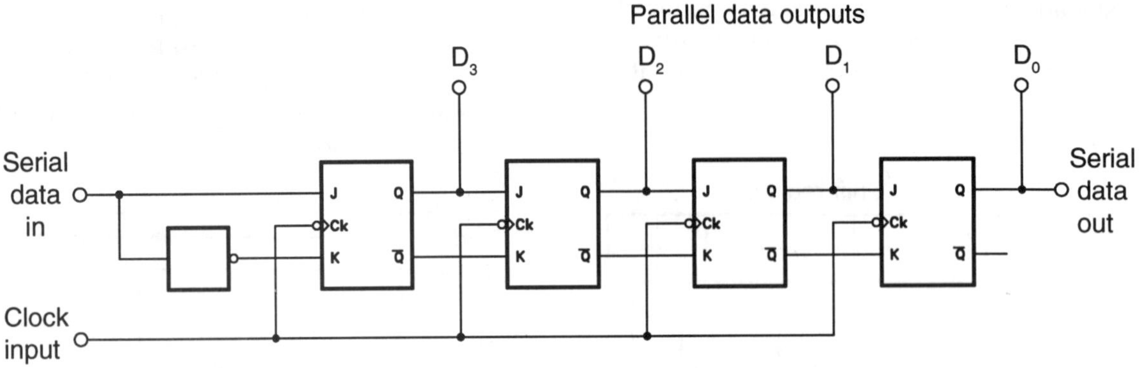

Parallel data outputs

D_3 D_2 D_1 D_0

Serial data in

Serial data out

Clock input

Method

a) Construct the above circuit.

b) By applying clock pulses (from a pulse generator or low frequency oscillator), complete the table opposite.

(The bistables used here are negative edge triggered types, so the Q output changes on a 1 to 0 transition of the clock.)

Output State Sequence

Clock	Serial In	D_3	D_2	D_1	D_0
1→0	1	0	0	0	0
1→0	0				
1→0	1				
1→0	1				
1→0	1				
1→0	0				
1→0	0				
1→0	0				

Practical Exercise   Running Light Displays

Introduction

If a series of LEDs are connected to an output port, attractive 'running light displays' may easily be created using shift and rotate instructions.

This is also a good way to learn about the many available types of shift and rotate instruction.

Method

a) Use shift left and shift right instructions to move a single illuminated light from left to right and back again, in a repeating sequence.

b) Use a rotate instruction to create a moving pattern that constantly moves in one direction.

If the accumulator initially contains 87_{16} and the carry flag is set, successive shift right (logical) instructions cause each bit to move to the right, as shown in table 4.7

	Accumulator	Carry
Initial values	1 0 0 0 0 1 1 [1]	1
After 1 shift	0 1 0 0 0 0 1 |1|	1
After 2 shifts	0 0 1 0 0 0 0 |1|	1
After 3 shifts	0 0 0 1 0 0 0 |0|	1
After 4 shifts	0 0 0 0 1 0 0 |0|	0
After 5 shifts	0 0 0 0 0 1 0 |0|	0
After 6 shifts	0 0 0 0 0 0 1 |0|	0
After 7 shifts	0 0 0 0 0 0 0 [1]	0
After 8 shifts	0 0 0 0 0 0 0 0	1
After 9 shifts	0 0 0 0 0 0 0 0	0

Table 4.7 Shift right (logical) example.

By comparing the least significant bit of the accumulator after each shift instruction (shown highlighted) with the original accumulator content, it can be seen that this instruction converts parallel data into serial form. The only additional requirement to enable serial data transfer is that the LSB of the accumulator must be sent to an output port line after each shift.

Similar techniques may be used to convert incoming serial data into parallel form. In this case, a *rotate* instruction may be used to reconstruct the parallel data.

If the serial data is transmitted LSB-first (as in the above example), the input data may be shifted first into the carry flag, and then rotated into the destination register, as shown in figure 4.21.

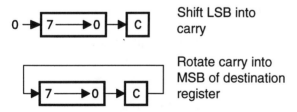

Shift LSB into carry

Rotate carry into MSB of destination register

Fig. 4.21 Converting serial data to parallel.

This process must be repeated eight times in order to reconstruct the parallel data, as shown in table 4.8.

	Data Register	Carry
Initial content	0 0 0 0 0 0 0 0	0
After 1 shift		1
After 1 rotate	1 0 0 0 0 0 0 0	0
After 2 shifts		1
After 2 rotates	1 1 0 0 0 0 0 0	0
After 3 shifts		1
After 3 rotates	1 1 1 0 0 0 0 0	0
After 4 shifts		0
After 4 rotates	0 1 1 1 0 0 0 0	0
After 5 shifts		0
After 5 rotates	0 0 1 1 1 0 0 0	0
After 6 shifts		0
After 6 rotates	0 0 0 1 1 1 0 0	0
After 7 shifts		0
After 7 rotates	0 0 0 0 1 1 1 0	0
After 8 shifts		1
After 8 rotates	1 0 0 0 0 1 1 1	0

Table 4.8 Converting serial data to parallel using the rotate right instruction.

If the final content of the data register is examined, it can be seen to contain 87_{16}, which was the serial data transmitted in the earlier example.

(This explanation neglects the generation and checking of start, stop and parity bits, which must also be performed by the software.)

In general, a serial interface implemented in software is likely to be used in systems where component costs (and numbers) must be minimised. A heavy load is placed on microprocessor time by such software intensive arrangements.

Practical Exercise   Software Serial Interface

Introduction

This exercise demonstrates the connection of two computers using a serial interface which is implemented as one bit of a parallel port.

Software techniques are used to perform parallel to serial conversion at the transmitter and serial to parallel conversion at the receiver.

Method

a) Write software on the transmitter computer which will wait for a key to be pressed by the user. The ASCII code of the pressed key should then be transmitted across the serial interface.

b) Write software on the receiving computer which will wait for incoming serial data. The program should convert the serial data into parallel form before displaying it on the screen.

(Both programs should run continuously.)

Conclusions

a) Use an oscilloscope to record typical data transfer waveforms and relate these to the ASCII codes transmitted.

b) Does the use of TTL logic levels limit the maximum cable length? Why?

c) How could your system be adapted to allow bi-directional data transfer?

Computer Connections

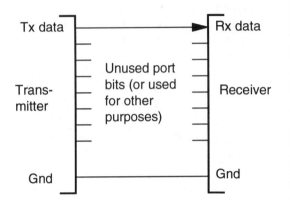

Suggested Z80 Techniques

For the receiver...

Register	Purpose
A	Input data (LSB)
B	Loop counter
C	Data register

Suggested 6502 Techniques

For the receiver..

Register	Purpose
A	Input data (LSB)
X	Loop counter
Memory location	Data register (RAM)

RS232 Serial Interface

The RS232 interface is widely used in serial data communication, and its main features are outlined here.

This interface was originally intended to specify the electrical connections between a *terminal* and a *modem*. The terminal was referred to as the *data terminal equipment* or *DTE*, while the modem was known as the *data communication equipment* or *DCE*.

In this context, a terminal is an arrangement of a keyboard and screen used for data communications purposes. Characters typed on the keyboard are transmitted via the serial port, while incoming serial data is displayed on the screen. (It is more common nowadays to simulate the function of a terminal by running a *terminal emulation* program on a general purpose computer.)

A modem is used to convert binary signals to and from audio frequency waveforms which may be transmitted directly across a telephone line. A commonly used technique, known as *frequency shift keying* (FSK), uses two different audio frequencies to represent the allowed binary logic levels. Circuitry in the receiving modem then converts these audio waveforms back into digital form. The term 'modem' is a shortened form of *modulator/demodulator*, which aptly describes its function.

Figure 4.22 illustrates the required equipment for two way computer communication using the telephone network.

It can be seen that each transmitted character passes through several stages as it moves from the transmitter to the receiver. These include

- parallel to serial conversion at the transmitter (signals are digital here),
- modulation (signals become audio frequency analogue waveforms),
- transmission across the telephone network in analogue form,
- demodulation (signals are converted back to serial digital waveforms),
- serial to parallel conversion. (Parallel digital data can be directly interpreted by the microprocessor).

Now consider the detail of the RS232 interface specification itself.

The first surprise is that TTL logic levels are not used for the transmitted and received data. Instead, a logic 1 (at the transmitter) is defined as being between –5 and –25 volts, while the logic 0 level is represented by a voltage in the range +5 to +25 volts. At the receiver, any voltage greater than +3 volts is interpreted as a logic 0, while voltages less than –3 volts are recognised as logic 1. This gap of 6 volts between the two logic levels ensures a much improved noise immunity for RS232 signals, when compared with TTL.

The use of non-standard voltage levels by the RS232 interface results in added complexity, including the need for additional power supply voltages and special circuitry to convert data signals to and from RS232 voltage levels.

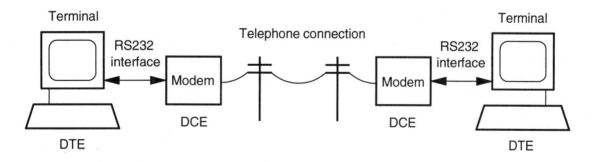

Fig. 4.22 Typical RS232 application and terminology.

Practical Exercise   RS232 Investigation

Introduction

This exercise examines the transmission of serial data using the RS232 interface.

Method

a) Connect two terminals (or two computers running terminal emulation software) via a suitable RS232 cable.

b) Ensure that both computers are set to the same communication settings (baud rate, parity, number of stop bits etc.).

c) Confirm that data typed on the keyboard of one terminal is displayed on the screen of the other.

d) Use an oscilloscope to record the actual waveforms transmitted for typical ASCII characters (ASCII 'U' is a good character to use when setting up the oscilloscope, due to its alternating bit pattern). Hence complete the table opposite.

Conclusions

a) What voltages are used by the RS232 interface to represent binary logic 0 and logic 1? How are these signals converted to and from TTL logic levels?

b) By comparing each ASCII code with the actual transmitted data, explain the significance of each section.

Computer Connections

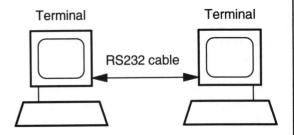

Terminal   Terminal

RS232 cable

RS232 Cable Wiring

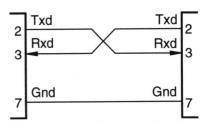

Observed Waveforms

Character	ASCII code	Actual bit pattern
A	100 0001	
B	100 0010	
C	100 0011	
D	100 0100	
E	100 0101	
F	100 0110	
G	100 0111	
H	100 1000	
I	100 1001	
J	100 1010	
K	100 1011	
L	100 1100	
M	100 1101	
N	100 1110	

(The need for additional power supplies is overcome in some RS232 driver/receiver ICs, which use *charge pump* techniques to generate the required RS232 voltage levels from the normal TTL supplies.)

The RS232 standard defines 25 different pin functions, although not all of these are in common use. Table 4.9 lists the most important signals, together with their main function.

Pin	Name	Function
1	GND	Chassis ground (connected to the metal case, if any).
2	TXD	Transmitted data (output data from the DTE or input data to the DCE).
3	RXD	Received data (output data from the DCE or input data to the DTE).
4	RTS	Request to send. An output from the DTE indicating its readiness to transmit data.
5	CTS	Clear to send. An output from the DCE, which permits data transmission.
6	DSR	Data set ready. An output from the DCE which indicates that it is operational (powered and on-line).
7	SG	Signal ground (establishes logic levels at both ends of the interface).
8	DCD	Data carrier detect. An output from the DCE indicating that a carrier signal has been detected (from the other modem).
20	DTR	Data terminal ready. An output from the DTE which indicates that it is operational (powered and on-line).
22	RI	Ring indicator (indicates that the telephone is ringing to a modem with an auto answer facility).

Table 4.9 Commonly used RS232 signals.

If the previous table is examined, it can be seen that two data lines (TXD and RXD) are used, allowing data transfer in both directions. Some interfaces are only capable of transferring data in one direction at any time, which is known as *half duplex* operation. Others allow simultaneous bi-directional operation, which is called *full duplex*.

The simplest possible RS232 data link consists of only three wires. These are the TXD, RXD and SG (signal ground) connections, as shown in figure 4.23.

DTE to DCE...

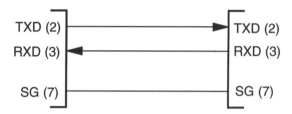

DTE to DTE (or DCE to DCE)...

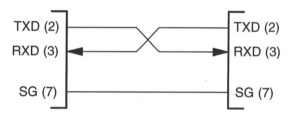

Fig. 4.23 Simple RS232 cables.

The first cable consists of direct connections between pins 2, 3 and 7 on the DTE and the same pins on the DCE.

In cases where two computers need to be linked via a serial interface, these are both likely to be DTEs (or DCEs). This means that pin 2 at one end must be linked to pin 3 at the other end (and vice versa), to allow proper connection. This arrangement is often called a *null modem* cable.

In addition to the simple arrangement shown above, it may be necessary to provide connections to other RS232 signals to allow reliable data transfer. A typical null modem cable, including these additional connections, is shown in figure 4.24 overleaf.

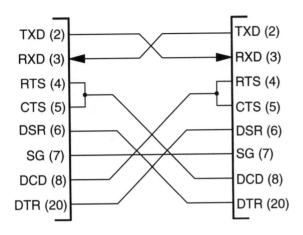

Fig. 4.24 Null modem RS232 cable.

By comparing the above connections with the pin descriptions, it can be seen that these signals fall into pairs such as

- TXD and RXD,
- DTR and DSR,
- RTS and CTS.

As can be seen, the connections used by the null modem cable are used to 'fool' the computers at each end that they are connected to a modem!

A final aspect of the RS232 interface is *flow control*, which is the control of the flow of information across the interface. An example would be the connection of a computer to a printer via a serial interface, where the data flow must pause temporarily if the printer buffer begins to fill (more than 90% full for example).

A potential problem with the use of a printer buffer would be the loss of data that would occur if characters continued to be sent, once the buffer had completely filled. Flow control techniques may be used here to start and stop data transfer, as required.

In software (*XON/XOFF*) flow control, two special ASCII codes are to control data transfer. These are 13_{16} (*DCE3* or *CTRL + S*) and 11_{16} (*DCE1* or *CTRL + Q*), which are used to pause and restart the interface respectively. Some manufacturers also use hardware (DTR or CTS) flow control, although this is not part of the RS232 standard.

Counting and Time Measurement using Hardware

A range of hardware devices are available which allow

- measurement of time intervals,
- event counting,
- determination of the current date and time.

Time intervals can be generated in software, in the form of software time delay routines (which were considered in the first volume), although this is very software intensive. A more flexible approach is to program a hardware device to generate an interrupt request once the required time interval has elapsed. This approach allows the microprocessor to perform other tasks while awaiting the interrupt.

An *interval timer* IC is based on a binary *down counter*, which uses additional logic to generate an interrupt once the count reaches zero. A simple interval timer is shown in figure 4.25.

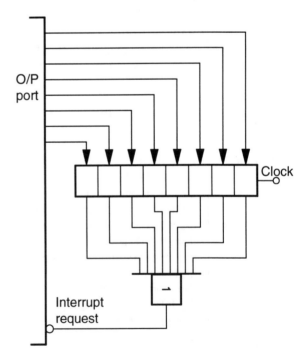

Fig. 4.25 Binary counter as an interval timer.

The time interval generated by this circuit may be changed either by altering the initial value loaded into the counter, or by altering the pulse repetition frequency of the clock input. Both of these techniques may be used with commercially available interval timer ICs.

Many interval timers may be programmed to produce a repeating stream of interrupts, rather than measuring a single time interval. In this case, once the counter has decremented to zero, it is automatically reloaded with its initial value from a *reload register*. This facility is useful in cases where the microprocessor must perform some regular (interrupt driven) action, in addition to its normal duties.

If the clock input is connected to an external signal, then the circuit of figure 4.25 may be used to count external events. In this case, each clock pulse might represent a component passing along a production line (breaking an infrared beam or passing a proximity sensor for example). By loading an initial value into the counter, it then becomes possible to generate an interrupt request once a certain number of external events has occurred. Alternatively the microprocessor could determine the number of events that had occurred at any time, simply by reading the state of the counter outputs.

As well as measuring time intervals or counting external events, it is also useful to be able to determine the actual time and date.

This facility is used by some operating systems to mark data or program files with their creation date and time. Date and time stamping of files provides useful information to computer users.

Real time clock ICs consist of a cascaded series of counters, each of which is incremented by the output of the previous stage. The current state of each counter is available in a register which may be read or altered by the user. Available registers typically include

- seconds (and fractions of a second),
- minutes,
- hours,
- day (of month and/or of week),
- month,
- year (including leap years).

Real time clock ICs are normally fitted with a battery-backed power supply, which enables them to maintain the correct time, even when the computer is switched off. This also eliminates the need to set the date and time manually, each time the computer is switched-on.

Mechanisms must also be provided to ensure the contents of internal registers are not updated mid-way through a read operation. In addition, a certain degree of 'intelligence' must be built into the real time clock IC, to enable it to cope with peculiarities of the calendar, including the variable length of each month, and leap years!

Practical Exercise   Software Real Time Clock

Method

a) Connect a 1 Hz squarewave to the non maskable interrupt pin.

b) Write interrupt driven software that updates the content of *seconds*, *minutes*, and *hours* 'registers', following each interrupt. (The valid range for *seconds* and *minutes* is 0 to 59, with 0 to 23 for *hours*.) Store data in BCD form.

c) Write subroutines to set or read or set any of the real time clock registers.

Hint

Use a software 'busy flag' to prevent unwanted updating of RTC registers by the ISR, during a read or write operation.

Z80 Counter/timer IC

The Z80 counter/timer IC provides four 8 bit counter/timers. Using this IC it is possible to produce single or repeating interrupt requests based on an elapsed time interval, or count external events. A block diagram of this IC is shown in figure 4.26.

From a programmer's point of view, the counter/timer appears as a contiguous group of four control registers. Each control register is used to select the detailed operation of its associated counter/timer channel.

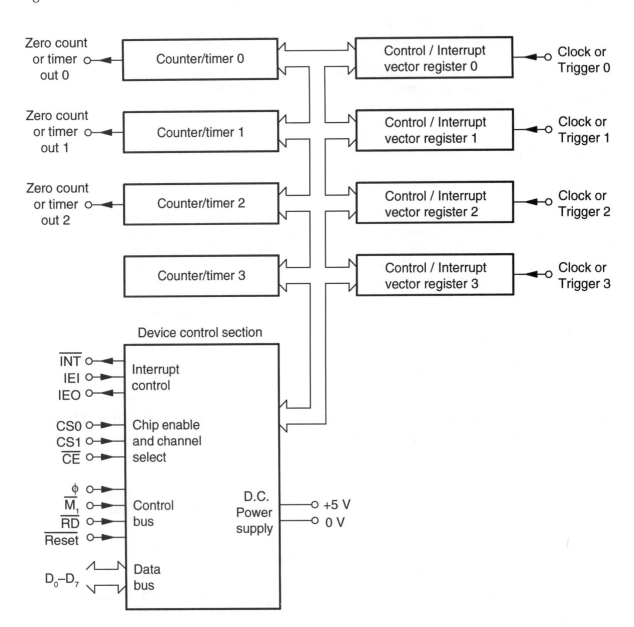

Fig. 4.26   Z80 counter/timer IC block diagram.

Each of the four channels has an external clock or trigger input. This signal is used to decrement the counter when the latter is in event counting mode. When operating in timer mode, a pulse of the correct polarity may optionally be used to begin the time interval. Alternatively, the timer may be programmed to commence automatically, once it is initialised.

Each channel, with the exception of channel 3, has an output pin which is used to indicate a *zero count* (the counter has decremented to zero), or *timer out* (the time interval has elapsed) condition. (The omission of an output for channel 3 is due to a shortage of pins on the counter/timer IC.)

The function of the control register depends upon the state of the least significant bit of the register. When this bit is zero, the remaining bits are used to specify the base address of an *interrupt vector table entry*. (Recall that Z80 peripherals are designed to operate in interrupt mode 2 which allows the use of vectored interrupts and daisy chain interrupt priority.) When the least significant bit of the control register is set to one, the remaining bits are used for general configuration purposes. Figure 4.27 shows the significance of each bit of a typical control register.

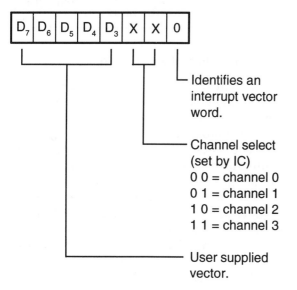

Fig 4.27 (a)   Interrupt vector byte.

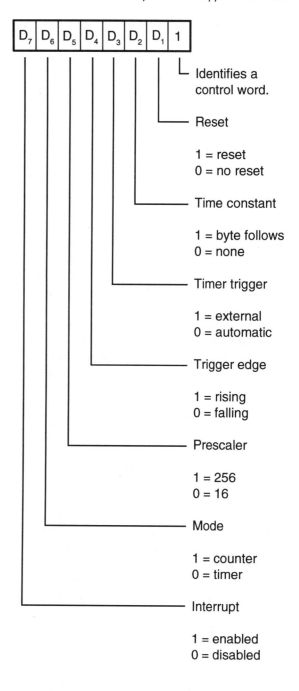

Fig. 4.27 (b)   Control byte.

Consider the general configuration of a channel, (assuming the LSB of the control byte is set to one).

The counter/timer IC may be reset externally by an active low $\overline{\text{RESET}}$ signal. Alternatively, individual channels may be reset in software by setting bit D_1 of the control register. This has the effect of disabling any count that is in progress, and also disables any pending interrupt request.

The *timer trigger* and *prescaler* bits are only relevant when the channel is programmed to operate in timer mode. A *time constant* byte, which is programmed immediately after a control word (assuming that bit D_2 is set), specifies the initial value that is loaded into the counter. This value is repeatedly decremented until the counter reaches zero, at which point an interrupt request may optionally be generated. The counter is then automatically reloaded with the time constant value, causing the counter to run continuously until reset.

Setting the timer trigger bit causes the timer to commence immediately following initialisation. When cleared, this bit causes the timer to wait for a pulse of the correct polarity (as determined by the trigger edge bit), before commencing.

In timer mode, the counter is decremented by the system clock, which is divided by a *prescaler* value (set to either 16 or 256 by bit D_5). The actual time interval generated by the timer may easily be calculated, based on a knowledge of the system clock frequency, prescaler and time constant values. For example, if the Z80 clock is 4 MHz and the prescaler value is set to 256, then the counter will be decremented once every 64 µs.

Counter operation (selected by setting bit D_6) also makes use of the time constant setting to specify the initial value placed into the counter/timer register. In this case, the counter is decremented each time a pulse of the correct polarity (programmed using bit D_4) is applied to the trigger input. Once again, an interrupt request may optionally be generated when the counter reaches zero.

Assuming that interrupts are enabled, it is also necessary to program an interrupt vector byte, which is identified by the LSB of the control register being cleared to zero.

This interrupt vector byte is combined with the I register (inside the Z80) to form the base address of the interrupt vector table, which in turn contains a series of pointers to each of the four interrupt service routines.. The structure of the interrupt vector table is shown in figure 4.28.

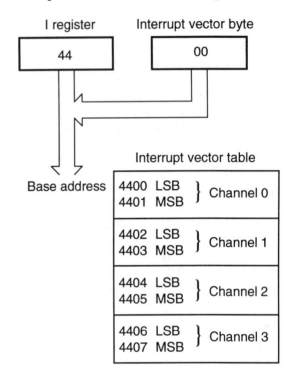

Fig. 4.28 Interrupt vector table.

Bits D_2 and D_1 of the interrupt vector byte are automatically set by the counter/timer, which means that only the most significant five bits are set by the programmer. This selects a group of eight addresses which are assumed to contain the address of each ISR (stored low byte first). Due to the automatic setting of these two bits, it is only necessary to write an interrupt vector byte for a single channel, in order to define the start address of the interrupt vector table.

As an example, assume that the interrupt vector table is stored in memory at address 4400_{16}, and the ISRs for channels 0, 1, 2, and 3 are stored at locations $4F00_{16}$, $4F40_{16}$, $4F80_{16}$ and $4FC0_{16}$ respectively. Listing 4.11 shows the necessary software configuration, and illustrates the production of a repeating timer interrupt.

```
        ORG  4000H
PIO: EQU  0          ;Base add. of PIO
DRA: EQU  PIO        ;Data register A
DRB: EQU  PIO + 1    ;Data register B
CRA: EQU  PIO + 2    ;Control register A
CRB: EQU  PIO + 3    ;Control register B

CTC: EQU  10H        ;Base addr. of CTC
CR0: EQU  CTC        ;Channel 0
CR1: EQU  CTC + 1    ;Channel 1
CR2: EQU  CTC + 2    ;Channel 2
CR3: EQU  CTC + 3    ;Channel 3

        EI           ;Enable interrupts
        IM  2        ;in Mode 2
        LD  A,44H    ;MSB of interrupt
        LD  I,A      ;vector table = 44H
        LD  A,0      ;LSB of interrupt
        OUT (CR0),A  ;vector table = 00H
        LD  A,A5H    ;1010 0101
        OUT (CR0),A  ;Select timer mode
        LD  A,80H  ; ;Time constant=80H
        OUT (CR0),A
        LD  A,0FH    ;Byte output
        OUT (CRA),A  ;on port A
        OUT (CRB),A  ;and port B
        LD  A,0
        OUT (DRB),A  ;Set ports to zero
LP:     OUT (DRA),A
        INC A
        JP  LP       ;Count up on port A

        ORG  4400H   ;Interrupt vectors
        DEFW 4F00H   ;Channel 0 pointer
        DEFW 4F40H   ;Channel 1 pointer
        DEFW 4F80H   ;Channel 2 pointer
        DEFW 4FC0H   ;Channel 3 pointer

        ORG  4F00H   ;Channel 0 ISR
        PUSH AF      ;Push registers
        IN  A,(DRB)  ;Read port B
        INC A        ;Increment count
        OUT (DRB),A  ;Write to port B
        POP AF       ;Pop registers
        EI           ;Enable interrupts
        RETI         ;Return from INT
```

```
        ORG  4F40H   ;Channel 1 ISR
        RETI

        ORG  4F80H   ;Channel 2 ISR
        RETI

        ORG  4FC0H   ;Channel 3 ISR
        RETI
```

Listing 4.11 Example counter/timer program.

This program produces an incrementing count on port A, which is generated by the main program. In addition, the counter/timer IC has been initialised to generate a repeating sequence of interrupts, by programming channel 0 as a timer. The counter/timer interrupt causes port B to be incremented. Thus, port A is incremented by the main program, while port B is incremented by the interrupt service routine.

The timer mode is set by loading $A5_{16}$ into the control register for channel 0. This selects the following options.

- Enable interrupts ($D_7 = 1$).
- Select timer mode ($D_6 = 0$).
- Prescaler = 256 ($D_5 = 1$).
- Trigger edge falling (unused) ($D_4 = 0$).
- Timer trigger set to automatic ($D_3 = 0$).
- Time constant byte follows ($D_2 = 1$).
- No reset ($D_1 = 0$).
- This is a control word ($D_0 = 1$).

Notice that due to the setting of bit D_2, a time constant byte must be written to the control register immediately following the control byte. In this example a value of 80_{16} is used for the time constant. Surprisingly, the maximum delay is generated by specifying 00_{16} as the time constant, as this is interpreted as 256_{10}!

On reaching zero, the timer is automatically reloaded with the time constant value, which causes a continuous stream of interrupts to be generated (until the IC is reset or reprogrammed).

Practical Exercise   Digital Frequency Meter

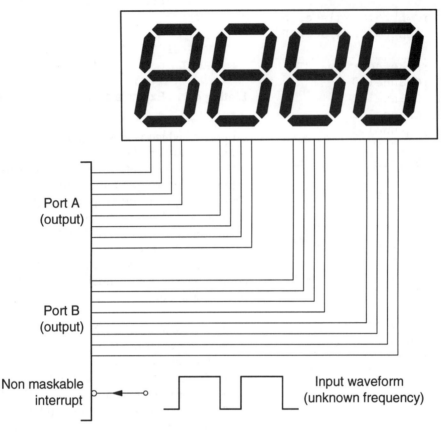

Four digit 7 segment display.

(Other types of display could also be used here)

Port A
(output)

Port B
(output)

Non maskable interrupt

Input waveform
(unknown frequency)

Introduction

This exercise demonstrates the practical use of a counter/timer IC to measure time intervals or count external events.

Method

a) Use a timer IC to generate interrupt requests at regular intervals of time.
b) Write software which will use the above timer interrupts to measure accurately a one second time interval.

c) Connect an externally generated digital waveform to the non maskable interrupt pin and write software which increments an internal counter after each non maskable interrupt.
d) Once the one second time interval has elapsed, send the current count (the signal frequency) to the display.
e) The program should run continuously.

Using the 6522 as a Counter/timer

The 6522 VIA (which was introduced previously in this chapter), is also fitted with a pair of 16 bit timers known as *timer 1* and *timer 2*. As shown earlier (in table 4.5), a total of six of the 6522's memory mapped registers are devoted to these timers. The first four registers, from offset 4 onwards, are devoted to timer 1, while the remaining two registers are used by timer 2. Clearly, from its greater number of registers, timer 1 is the more complex of the two!

Timer operation is configured using the most significant three bits of the *auxiliary control register* (*ACR*), as shown in figure 4.29.

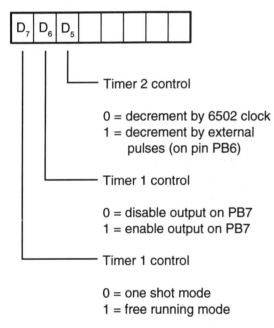

Fig. 4.29 Auxiliary control register timer control functions.

As can be seen, timer 2 may be decremented either by the system clock, or by pulses applied to line 6 of port B (in which case this external pin is no longer available as a general purpose input/output port bit).

In either case, only a single interval is generated. Once the 16 bit timer 2 register has decremented to zero, this causes the appropriate bit in the *interrupt flags register* (*IFR*) to be set, and may optionally generate an interrupt request.

Figure 4.30 shows the section of the IFR which is relevant to timer operation.

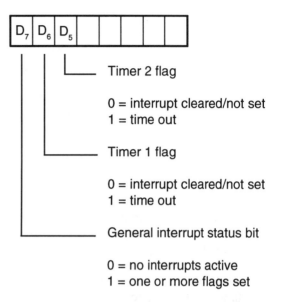

Fig. 4.30 Interrupt flag register timer functions.

A flag may be cleared by (as mentioned previously) by writing a 1 into the appropriate bit of the IFR. (The timer 2 interrupt flag may also be cleared by a read operation from the timer 2 counter/latch register, at offset 8.)

Each flag in the IFR corresponds with an identical bit in the IER. It is the status of the appropriate IER flag that determines whether an interrupt request is generated on completion of the selected counter/timer operation.

The following example demonstrates the generation of a simple time delay using timer 2.

```
BASE:   EQU $8000    ;6522 setup
T2C-L:  EQU BASE+8   ;Alter as req'd
T2C-H:  EQU BASE+9
ACR:    EQU BASE+11
IFR:    EQU BASE+13
        ORG $4000    ;Main program
        LDA ACR      ;Get existing ACR
        AND #$DF      ;1101 1111
        STA ACR      ;Clear T2 mode bit
```

Continued overleaf...

```
        LDA #0
        STA T2C-L     ;Clear T2 low byte
        LDA #2        ;Count = 512 cycles
        STA T2C-H     ;Start timer
LP:     LDA IFR       ;Wait for timeout
        AND #$20      ;Test T2 flag
        BNE LP        ;Wait until set

        LDA T2C-L
                      ;Clear IFR flag by
                      ;reading T2C-L
                      ;Put rest of main
                      ;program here...
```

Listing 4.12 Simple time interval using T2.

In this case timer 2 is loaded with an initial value of 512_{10}, which is then decremented to zero by the system clock. A polled approach is used to test the state of the appropriate flag in the IFR, at which point the main program continues.

A few simple modifications to the above program allow it to be used to count external pulses on data line PB6, as shown in listing 4.13.

```
BASE:   EQU $8000     ;6522 setup
DDRB:   EQU BASE+2    ;Alter as req'd
T2C-L:  EQU BASE+8
T2C-H:  EQU BASE+9
ACR:    EQU BASE+11
IFR:    EQU BASE+13
        ORG $4000     ;Main program
        LDA #0        ;Set port B as
        STA DDRB      ;an input port
        LDA ACR       ;Get existing ACR
        ORA #$20      ;0010 0000
        STA ACR       ;Set T2 mode bit
        LDA #$28      ;LSB = 40 (decimal)
        STA T2C-L     ;Clear T2 low byte
        LDA #0        ;MSB = 0 (total=40)
        STA T2C-H     ;Start counter
LP:     LDA IFR       ;Wait for count
        AND #$20      ;Test T2 flag
        BNE LP        ;Wait until set
        LDA T2C-L     ;Clear IFR flag by
                      ;reading T2C-L
                      ;Put rest of main
                      ;program here...
```

The operation of timer 1 is controlled by bits D_7 and D_6 of the ACR. Bit D_7 selects either free running (continuous) or single shot operation, while D_6 enables or disables an output on data line PB7. This can be a single pulse in the case of single shot mode or a change in state in free running mode, following each timeout. Using this feature it is possible to produce a repeating square waveform on PB7, whose frequency may be determined by a simple software configuration. This is useful in many applications, including serial interface baud rate generation.

A total of four 6522 registers are devoted to timer 1. A 16 bit latch at offsets 6 and 7 is used to reload the timer register (at offsets 8 and 9) automatically once a zero count has been reached, thus permitting free running timer operation. The following program demonstrates the use of timer 1 to produce a repeating square waveform signal on line PB7.

```
BASE:   EQU $8000     ;6522 setup
DDRB:   EQU BASE+2    ;Alter as req'd
T1L-L:  EQU BASE+6
T1L-H:  EQU BASE+7
ACR:    EQU BASE+11
        ORG $4000     ;Main program
        LDA #$FF      ;Set port B as
        STA DDRB      ;an output port
        LDA ACR       ;Get existing ACR
        ORA #$C0      ;1100 0000
        STA ACR       ;Set T1 mode bits
        LDA #$28      ;LSB = 0
        STA T1L-L     ;Clear T1 low byte
        LDA #4        ;1024 clock cycles
        STA T1L-H     ;Start timer

                      ;Put rest of main
                      ;program here...
```

Listing 4.14 Squarewave generation on PB7.

In this example, PB7 is low for 1024 clock cycles, and then high for the same interval in a repeating sequence.

Alternately, the PB7 output could be disabled. The above program could be then be modified to call an interrupt service routine at regular intervals.

Summary

The main points covered in this chapter are:

- A *parallel* interface may be used to link a computer to a peripheral such as a printer.
- *Hanshake lines* are often used to control the flow of information across such interfaces, and to indicate device status.
- The *Centronics* interface is widely used for parallel printer connection to computers.
- Typical PIO and VIA ICs may be programmed to transfer parallel data using handshaking and interrupts.
- Long distance data communication is best performed in *serial* form. Advantages of serial techniques include a reduction in cabling costs, greater noise immunity and the possibility of simultaneous bi-directional data communication.
- Serial data may be converted to and from parallel form using a *shift register*, or using special software routines.
- The actual serial data is enclosed or *framed* by one or more *start* and *stop bits*, which are used for timing and synchronisation purposes.
- An odd or even *parity* bit may optionally be added to the serial data to allow error detection (but not correction).
- Serial data may be transmitted across a telephone line by the use of a *modem* (a contraction of modulator/demodulator), which transforms the binary data to and from audio frequency signals.
- The *RS232* interface is widely used for serial data transfer, and was originally developed to define computer to modem connections.
- *Counter/timer* ICs may be used to count external events or generate predetermined time intervals. These may be single or repeating intervals, and may optionally be used to generate interrupt requests.
- *Real time clock ICs* provide access to date and time information. Battery-backed operation allows these devices to maintain their internal data, even when the computer is switched off.

Problems

1 Explain the method of operation and the advantages of using handshaking when transferring data between a computer and a peripheral.

2 State the main features of the Centronics parallel interface.

3 Compare the advantages and disadvantages of serial and parallel data communication.

4 Show how a shift register may be used to convert data from serial to parallel form and vice versa.

5 Outline the main features of the RS232 serial interface.

6 Explain the use of parity generation and checking in serial data communication.

7 Define the following terms.

   a) Start bit.
   b) Stop bit.
   c) Even parity.
   d) Null modem cable.

8 Explain the operation and use of a typical real time clock IC.

9 State the three main uses of a counter/timer IC.

5   Trends in Microprocessor Design

Aims

When you have completed this chapter you should be able to:

1 Understand the historical development of microprocessor based systems.
2 Appreciate current and future trends in microprocessor design.
3 Recognise shared characteristics of different microprocessor families, and the transferable nature of microelectronic design skills.

Historical Development

Following the invention of the transistor in 1947, the early 1950s saw a steady development in the electronics industry, with progressively more complex products being introduced. The transistor represented a great step forward in compactness and reliability over the previously dominant valve technology.

A second advance in the miniaturisation of electronic circuits occurred in 1959 with the introduction of the integrated circuit (IC). The 1960s then saw a rapid growth in the number of discrete components which could be fabricated onto a single slice of silicon. This figure doubled each year during this period.

By 1970, the first electronic calculators began to appear and research was underway to develop the first 'computer on a chip'. The first commercially available microprocessor was introduced by the Intel Corporation in 1971. This was a 4 bit device, commonly known as the Intel 4004, which was soon followed by the Intel 8008.

The complexity of microprocessors has increased steadily since their introduction, with recent designs containing more than 3 million transistors. At the same time, design improvements, such as increased operating speeds and memory addressing capacities have greatly improved system performance.

These trends are well illustrated by considering the development of the Intel family of microprocessors, which has expanded rapidly since the introduction of the original Intel 4004. Today, Intel microprocessors are found inside a high proportion of personal and business computers.

A year after the 8008, the much faster 8080 microprocessor was introduced. This in turn was superseded by the Intel 8085 which, due to its increased integration, required fewer external components. The 8085 offered software compatibility with its predecessor, combined with increased system performance, which was popular with both designers and users.

An interesting feature of the 8085 was the use of a *multiplexed* data bus arrangement. This feature (which was introduced due to restricted pin availability) meant that the lower half of the address bus and the 8 bit data bus shared the same pins, but at different times. An *address latch enable* (ALE) signal was used to latch the lower half of the address bus into an external latch IC. The output of this latch was then combined with the upper half of the address bus to form the full 16 bit address.

This 16 bit address bus, combined with the 8 bit data bus gave an addressing capacity of 64 kilobytes, which soon became a limitation as software complexity increased.

In 1978 Intel introduced the 16 bit 8086 microprocessor. The 8086 was designed to be software compatible with the earlier 8085 and 8080 designs, to encourage existing users to transfer to the newly introduced 16 bit systems. Users were further encouraged by the introduction of the Intel 8088, which had characteristics of both 8 bit and 16 bit systems. Externally, the 8088 possessed an 8 bit data bus and was compatible with existing 8085 peripherals, while internally it had 16 bit registers, making it software compatible with the 8086. The fact that 16 bit registers could only be transferred to or from memory one byte at a time, resulted in slower operation compared with the 8086, but the 8088 still proved popular. In fact, in 1981, IBM chose the Intel 8088 for its newly released *personal computer* – the *IBM PC*.

Members of the 8086 family could optionally be interfaced to an 8087 *maths coprocessor*, which allowed mathematical calculations to be performed at high speed in hardware. This chip had a range of *floating point* and integer-based functions, including trigonometric, logarithmic and general mathematical operations. The addition of a coprocessor greatly increased the speed of computationally intensive applications such as *spreadsheets* or *computer aided design* (*CAD*). An empty socket was provided with most new computers, making the coprocessor upgrade relatively straightforward.

The 20 address lines of the 8086/8 provided a 1 megabyte address space, which initially seemed large enough for any application that could be envisaged. However, as the complexity of software increased, it soon became a serious limitation.

In 1982 Intel released two new members to the rapidly expanding 80X86 microprocessor family. The highly integrated 80186 was intended for embedded control applications, while the 80286 was aimed at the *desktop* (a computer on every desk) and *workstation* (the computer as a design tool) markets. The 80286 had an increased address space of 16 megabytes, and also provided facilities for *multitasking* and *virtual memory*.

Multitasking, as its name suggests, allows the computer to give the appearance of running several programs (or tasks) simultaneously. In fact, each process runs for a fraction of a second, in a repeating cycle. At the end of each time interval, the execution state of the current process is saved, prior to control being passed to the next task in the sequence.

Virtual memory allows the microprocessor to execute programs which are much larger than the available physical memory. The result is that the computer appears to have more RAM than is actually the case. During program execution, only the currently active section of the program is loaded from backing store. If the microprocessor attempts to access an area of memory that is not currently available, this causes a special type of error called an *exception*. This in turn results in the required program segment being loaded from backing store, after which program execution continues normally. The only disadvantage of virtual memory is the slight reduction in operating speed caused by repeated disc accessing. This is compensated by the fact that program size is no longer limited by the availability of physical memory. Virtual memory is particularly useful in multitasking systems where the machine's resources are shared between several competing processes.

1985 saw the introduction of the Intel 80386 microprocessor, which was a considerable advance over the 80286. The 80386 was the first 32 bit microprocessor in the 80X86 family. In addition to the 32 bit data bus, the address bus was also increased to the same size, giving a direct addressing capacity of 4 GB (where 1 *gigabyte* = 1024 megabytes or 2^{30} bytes). Virtual memory facilities were also improved, giving a maximum program size of 64 *terabytes* (where 1 terabyte = 1024 gigabytes or 2^{40} bytes). As expected, the 80386 chip set also included a range of 32 bit peripherals, such as the 80387 maths coprocessor, direct memory access (DMA) and cache memory controller ICs.

In 1989, the Intel 80486 DX microprocessor was released, which offered a 100% improvement in operating speed (at the same clock frequency) over the previous 80386. The 80486 was the first member of the 80X86 family to contain in excess of one million transistors, and was fully compatible with previous generations. Performance increases in the 80486 were due partly to the adoption of electronic design principles pioneered in *reduced instruction set* or *RISC* microprocessors. These highly efficient RISC microprocessors achieve very high operating speeds, due to their simple instruction set, which is optimised to permit the majority of instructions to be completed in a single clock cycle.

At the time of writing, the first 64 bit microprocessors in the 80X86 family have started to appear. A break with tradition (for copyright protection purposes) has been the adoption of the trade name *Pentium*, rather than the expected 80586. The Pentium P24T is intended to be compatible with existing 486 motherboards (recall the 8088), while other Pentium microprocessors are recommended for new designs.

The Pentium microprocessor contains 3.1 million transistors and the silicon wafer has a surface area of almost 300 mm². Operating at up to 90 MHz, the Pentium executes instructions twice as fast as an 80486 with the same clock frequency.

Another technique adopted from RISC microprocessors has been the adoption of a *superscalar* architecture, which allows the microprocessor to execute several instructions in a single clock cycle. Normally, instructions and data are loaded from memory by the *bus interface unit (BIU)* and placed into an instruction queue, ready for subsequent execution by the *execution unit*. The BIU operates independently of the execution unit, and assumes that subsequent instructions will come from contiguous addresses. As each instruction in the instruction queue is executed, following instructions move up by one position. In the event of a jump, the queue is flushed and the BIU begins filling the queue from the new program address. This division of responsibility between the BIU and the execution unit results in efficient utilisation of memory and other resources.

Superscalar microprocessors possess multiple execution units, each of which can simultaneously execute a single instruction taken from the instruction queue. The microprocessor is able to identify groups of instructions that could be executed in any order (such as loading several numbers into different registers), and each of these instructions is passed to its own execution unit. In some superscalar microprocessors the ALU is capable of performing several simultaneous operations (such as a multiplication and an addition at the same time).

Each new generation of microprocessor has seen a steady increase in operating speed, with the latest (at the time of writing) 80486 DX/4 microprocessor operating at an effective clock speed of 100 MHz, and able to execute simple register transfer instructions in as little as 0.01 μs. This is a marked improvement over the 10 μs required for the same operation on the earlier Intel 8008 microprocessor! Thus in 20 years, the efficiency of the microprocessor has increased by at least 1000 times. If the efficiency of the motorcar had increased by the same amount, a typical tank full of petrol would last for 0.5 million miles (to the Moon and back!). Figure 5.1 illustrates the performance improvements achieved.

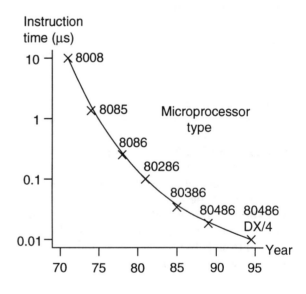

Fig. 5.1 Typical instruction times (μs).

Case Study – the Intel 8086

The following section introduces the Intel 8086 microprocessor, which is used here to illustrate the features of a typical 16 bit microprocessor. Readers who require a more detailed treatment of the 80X86 family of microprocessors should consult Intel's microprocessor databooks, which are updated annually, or any of the wide range of text books on the subject.

The 8086 microprocessor is divided into two sections, known as the *execution unit* and the *bus interface unit (BIU)*, as shown in figure 5.2.

As mentioned previously, the BIU is responsible for keeping the instruction queue as full as possible at all times, thus ensuring that the execution unit always has instructions and data to operate upon.

Notice also that the 20 address lines give a total memory addressing capacity of 1 megabyte, while the lower 16 lines of the address bus are time multiplexed with the 16 bit data bus. (This was also a feature of the 8085 microprocessor, mentioned earlier.)

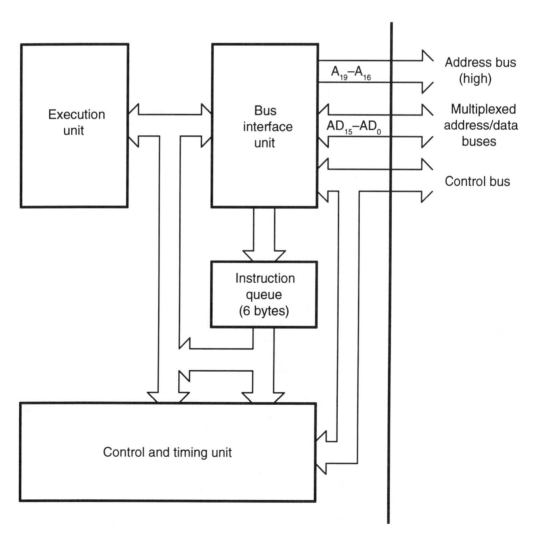

Fig. 5.2 8086 functional block diagram.

The use of multiplexing allows the Intel 8086 to be produced as a 40 pin package, as shown in figure 5.3.

1	GND	Vcc	40
2	AD14	AD15	39
3	AD13	A16/S3	38
4	AD12	A17/S4	37
5	AD11	A18/S5	36
6	AD10	A19/S6	35
7	AD9	$\overline{BHE}$/S7	34
8	AD8	MIN/$\overline{MAX}$	33
9	AD7	$\overline{RD}$	32
10	AD6	HOLD	31
11	AD5	HLDA	30
12	AD4	$\overline{WR}$	29
13	AD3	M/$\overline{IO}$	28
14	AD2	DT/$\overline{R}$	27
15	AD1	$\overline{DEN}$	26
16	AD0	ALE	25
17	NMI	$\overline{INTA}$	24
18	INTR	$\overline{TEST}$	23
19	CLK	READY	22
20	GND	RESET	21

Fig. 5.3 8086 (minimum mode) pin-out

The above illustration shows 8086 pin connections when the microprocessor is configured to operate in *minimum mode*, as selected by the logic level applied to pin 33. Minimum mode is normally used in smaller microcomputer designs, while *maximum mode* is intended for more complex applications. In maximum mode the function of certain pins are changed, which allows the microprocessor to operate with an 8288 *bus controller* IC. The 8288 then generates all external bus control and timing signals.

In either mode, one or more latch ICs are used to demultiplex the time multiplexed address/ data bus signals on pins 2–16 and 39. It is the falling edge of the ALE signal which is used to latch the lower 16 lines of the address bus, which occurs during T_1 of each machine cycle.

During the remainder of the machine cycle, lines AD_{15}–AD_0 act purely as a 16 bit data bus, while the lower 16 lines of the address bus are derived from the latch outputs. This is illustrated by figure 5.4.

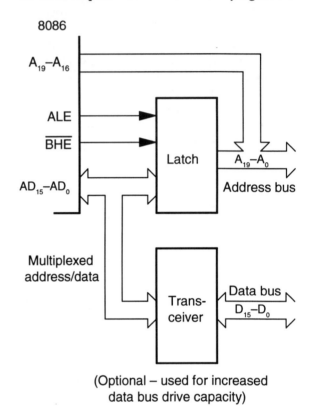

(Optional – used for increased data bus drive capacity)

Fig. 5.4 Address/data bus demultiplexing.

The memory in an 8086-based system is arranged in two banks, each of which is 8 bits wide. All odd numbered address requests are directed to one bank, while even addresses are referred to the other. The active low *bus high enable* pin ($\overline{BHE}$) indicates, when zero, that the upper half of the data bus (D_{15}/D_8) is being used to transfer data to or from an odd numbered address. Similarly, address line A_0 indicates, when low, that data is being transferred to or from an even numbered address, via the lower half of the data bus.

A 16 bit word may be transferred to or from memory by making $\overline{\text{BHE}}$ and A_0 simultaneously low. In this case, it is important that the number is placed on an even address boundary, if the number is to be transferred in a single machine cycle. Incorrectly aligned 16 bit data can still be accessed by the 8086, but two 8 bit memory accesses are then required, resulting in slower operation. (This restriction applies to data operands only, and not to instruction op. codes.)

8086 Memory Segmentation

The 8086 microprocessor can address up to 1 megabyte (1024 kB) of memory using its 20 address lines, with valid addresses in the range 00000–FFFFFH. As can be seen, the addressing capacity is 16 times larger than that previously encountered in 8 bit systems, thus requiring five hexadecimal digits to specify an absolute address.

Memory in an 8086-based system consists of several (possibly overlapping) segments, each of which is 64 kilobytes in size. A physical address in memory is referred to by combining the content of a 16 bit *segment register* with an *offset* value, as shown in figure 5.5.

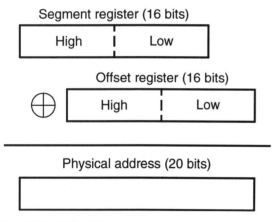

Fig. 5.5 8086 Memory segmentation.

In effect, the segment register content is shifted four places to the left before being added to the offset value. For example, if the segment register contained 5678_{16} and the offset value was 1234_{16}, the physical address referred to would be $579B4_{16}$ (which would normally be written as 5678:1234H).

The 8086 has four segment registers, each of which has a unique function.

● The *code segment* (*CS*) register is used as the base address when an instruction is fetched from memory, or when a jump occurs. The physical address of the next instruction is calculated by combining the CS register with the 16 bit *instruction pointer* (analogous to the program counter in an 8 bit system).

● Most operations which refer to the stack (such as subroutines, interrupt service routines and push and pop instructions) are based on the *stack segment* (*SS*) register. The physical address is produced by combining the stack segment and stack pointer registers. (The *base pointer* register also refers to the stack segment by default, as will be seen shortly.)

● The data segment (DS) register is used to refer to local program data and specifies the source memory range for *string* (block memory copy) operations. The physical address is produced by combining the data segment register with the general purpose register referred-to by the instruction.

● The extra segment (ES) provides an alternative data segment and is also used to refer to the destination memory range for string operations.

Although most 8086 instructions operate on a default segment register, this can in many cases be overridden.

For example, the instruction

```
MOV AX, [BP]
```

loads the AX register with 16 bit word pointed to by the base pointer register. By default the base pointer register refers to the stack segment, but this can be changed by using a *segment override* such as CS:, DS:, or ES:. To make the instruction refer to the extra segment for example, the instruction would be changed to

```
MOV AX, ES: [BP]
```

Segment registers allow the programmer to separate the program code from the data on which it operates, as shown in figure 5.6.

8086 Programmer Registers

Figure 5.7 shows the internal register arrangement used by the Intel 8086 microprocessor.

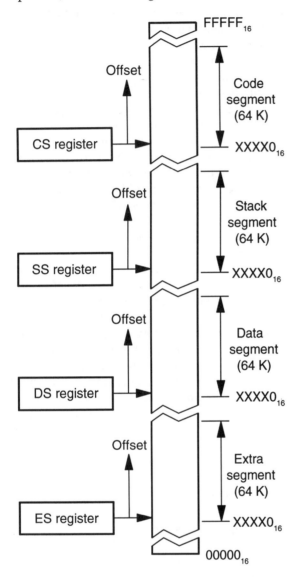

Fig. 5.6 Typical 8086 memory organisation.

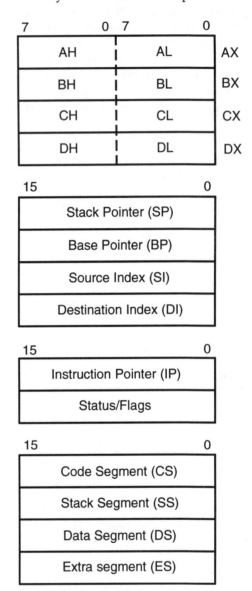

Fig. 5.7   8086 register model.

The segment register points to the base address of a particular segment, which is always on a 16 byte boundary. This is because the content of the segment register specifies the most significant 16 bits of the 20 bit physical address, while the least significant 4 bits of the segment address are forced to zero.

Most of the 8086 registers are associated with a particular segment register although, as mentioned previously, this can be overridden in many cases.

The *instruction pointer* register is analogous to the program counter in an 8 bit system, and is combined with the code segment register to form a 20 bit physical address.

Similarly the stack pointer performs the same function as its 8 bit counterpart. In this case the stack segment register points to the base address used by the stack, while the stack pointer provides the offset.

8086 status information is available by testing individual flags in the status/flags register. The arrangement of this register is shown in figure 5.8.

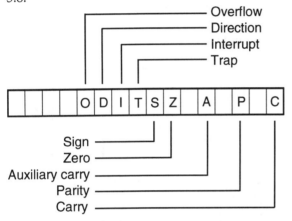

Fig. 5.8 8086 status/flags register layout.

Testing of particular flags is normally performed indirectly using conditional jump instructions. For example the instruction

```
JC LABEL      ;Jump if carry set
```

jumps to a new address if the carry flag has been set by a previous instruction.

The direction, interrupt and trap flags are used to control the operation of the microprocessor, rather than simply reflecting the internal state of the 8086. The direction flag is used by string operations (which will be explained shortly), while the interrupt flag enables or disables maskable interrupt requests. The trap flag is useful for software debugging as, when enabled, it causes a special software interrupt to be generated after each instruction is executed. This is normally used to implement a single step facility in monitor software.

General purpose 8086 registers include the AX, BX, CX and DX 16 bit registers. The high (most significant) or low (least significant) byte of these registers may also be referred to in an assembly language instruction, as shown in the following example.

```
MOV AX,1234H   ;Put 1234H into AX
MOV AH,12H     ;Repeat using two
MOV AL,34H     ;8 bit instructions
```

Note that the above instructions use *immediate addressing* to load 8 or 16 bit quantities into a register. It is the register name that specifies the size of the operand. Any register name ending in 'X' indicates a 16 bit quantity, while 'H' or 'L' refers to the high or low half of the appropriate data register. Negative values are stored in two's complement form, and are *sign extended* to fill the specified register size.

It is also possible to move data from one register to another (using *register addressing*), as shown below.

```
MOV AX,CX
```

The content of one or more memory locations may also be loaded into a data register using *direct addressing*. In this case the operand normally refers to a label which is specified elsewhere in the program. For example the instruction

```
MOV CL,TABLE
```

loads the content of a memory location into the low half of the CX register. This could also be written in the form

```
MOV CL,BYTE PTR TABLE
```

where confusion between immediate and direct addressing modes might occur. (*WORD PTR* would be used to specify a 16 bit operand stored at a pair of addresses.)

Although the data registers are defined as 'general purpose', they each have specific applications. The AX register is used as an accumulator by the majority of arithmetic instructions, while the BX or *base* register may be used to address data in memory. The CX or *count* register may be used as a loop counter by string copy instructions, while the DX register is used by multiplication and division operations. (The DX register is also used to specify a port address during some types of input/output operation.)

Example

Listing 5.1 demonstrates the addition of two 16 bit numbers using direct addressing.

```
MOV AX,WORD PTR 2000H  ;Get 1st No.
ADD AX,WORD PTR 2002H  ;Add 2nd No.
MOV 2004H,AX           ;Store result
```

Listing 5.1 16 bit addition using direct addressing.

(There is no possible confusion between immediate and direct addressing in the last instruction because the destination operand (on the left) cannot be an immediate value.)

Data is normally accessed from within the data segment, or in some cases from the extra segment. The offset is supplied by the BX register, or by either the source or destination index registers. For example the instruction

```
MOV AX, [BX]
```

uses the BX register (and *register indirect* addressing) to specify the offset in the data segment. This is a more flexible approach than direct addressing seen above, since a table of values in memory can be accessed in sequence, simply by incrementing or decrementing the appropriate register.

String instructions operate on blocks of data which may be up to 64 kilobytes in size. The data itself may consist of ASCII characters, packed or unpacked BCD digits or simply binary numbers. The data type is not important to the instruction.

Using string instructions a wide range of operation is possible.

- Copy a block of memory from one area to another.
- Compare two blocks of memory.
- Compare a memory block with the accumulator.
- Load or store data from the accumulator while automatically incrementing or decrementing the memory pointer.
- Input or output a memory block via a port.

As a typical example, consider the detailed operation of a string copy instruction, which is shown in figure 5.9.

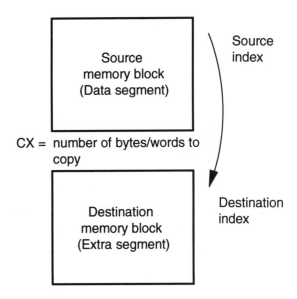

Fig. 5.9 String copy instruction.

The *source index* and *destination index* registers are first initialised to point to the source and destination memory ranges, while the CX (or count) register contains the number of characters to be copied. In this case a 'character' may be either a byte or a 16 bit word, depending on the exact copy instruction used (MOVSB moves a single byte, while MOVSW copies 16 bits at a time).

The source string is assumed to be in the data segment, while the destination string is placed in the extra segment by the copy instruction.

After each character is copied, the content of the CX register is decremented, while SI and DI are altered to point to the next memory location.

The state of the *direction flag* in the status register determines whether the SI and DI registers are incremented or decremented after each copy operation. If the direction flag is cleared, then SI and DI are automatically incremented following each copy, as shown in the following example.

Example

Write a program to copy 100_{16} bytes from offset 500_{16} in the data segment to offset 400_{16} in the extra segment.

The solution is given in listing 5.2 below.

```
CLD               ;Clear direction
                  ;flag
MOV CX,100H       ;Set byte count
LEA SI,FROM       ;Source in DS
LEA DI,ES:TO      ;Destination in
                  ;ES
REP  MOVSB TO,FROM ;Block copy
```

Listing 5.2 String copy program.

In this example it is assumed that the labels FROM and TO have been defined in the program, and that these are in the data and extra segments respectively. Due to the clearing of the direction flag, SI and DI are incremented after each byte is copied, so FROM and TO must refer to the lowest address in the source and destination memory ranges. The *load effective address* (LEA) instruction is used to place the appropriate offset values into the source and index registers. (Notice also the use of a segment override operator to refer to a label in the extra segment.)

The *repeat* prefix is placed in front of the MOVSB instruction to cause the instruction to repeat automatically until the CX register becomes zero. This removes the need for the programmer to explicitly place MOVSB within a program loop.

A major advantage of the 8086, when compared to a typical 8 bit microprocessor, is the inclusion of multiplication and division operations as part of the instruction set! This eliminates the need to develop special software algorithms for these functions (which were considered in *Microelectronic Systems – Book One*).

Two multiplication instructions are included within the 8086 instruction set. These are,

- MUL multiply unsigned integers,
- IMUL multiply signed (two's complement) integers.

The multiplicand and multiplier may be either bytes or 16 bit words. In each case, the product is twice as long as the original data.

For 8 bit multiplication the multiplicand is the content of the AL register, while the multiplier is specified by the multiply instruction's operand. The 16 bit result is stored in the AX register.

Example

Write a program to multiply 7×5 using the 8086 microprocessor.

```
MOV AL,7     ;Put multiplicand in AL
MOV BL,5     ;Put multiplier in BL
MUL BL       ;Unsigned multiply
MOV 10H,AX   ;Store result at offset
             ;10H (and 11H) in DS
```

With 16 bit multiplication, the multiplicand is placed in the AX register, with the 16 bit multiplier being provided by the instruction operand. The lower half of the 32 bit result is stored in the AX register (as for the 8 bit version), while the most significant word is placed in the DX register.

Note that the four basic mathematical operations can be carried out on packed or unpacked BCD data, as well as on signed or unsigned binary numbers. 'Correction' instructions are provided which convert the result to the valid data type, once the mathematical operation has been completed. These are similar in principle to the Z80's DAA or *decimal adjust accumulator* instruction.

Table 5.1 lists the BCD result correction instructions supported by the Intel 8086.

Instruction	Function
AAA	ASCII adjust for addition
DAA	Decimal adjust for addition
AAS	ASCII adjust for subtraction
DAS	Decimal adjust for subtract
AAM	ASCII adjust for multiplication
AAD	ASCII adjust for division

Table 5.1   BCD adjust accumulator operations.

None of these instructions require an operand to be specified. As a general rule, 'decimal adjust' instructions operate on packed BCD data while 'ASCII adjust' operations refer to unpacked BCD numbers. For example, the *ASCII adjust for multiplication* (AAM) instruction assumes that the multiplier and multiplicand were two unpacked BCD bytes. Following the AAM instruction, the most significant digit of the product is stored in the AH register (also in unpacked BCD form), while the least significant digit is stored in AL.

As with multiplication, the 8086 also provides a pair of divide instructions. The DIV instruction operates on unsigned integers, while IDIV is used for two's complement (signed) data. The instruction operand specifies the byte or word sized *divisor* (by which the *dividend* will be divided to produce the *quotient* and the *remainder*). Note that the divisor cannot be an immediate value so it must come from either a register or from memory.

The dividend is always twice as long as the divisor. For division by a byte-sized divisor, the 16 bit dividend is stored in the AX register. For 16 bit division, the 32 bit dividend is stored the DX and AX registers, with DX being the most significant half of the dividend.

8 bit division operations return the quotient in AL and the remainder in AH, while 16 bit division instructions return the quotient in AX and the remainder in DX.

Example

Write a program to divide a 32 bit double word (stored in the DX and AX registers), by the 16 bit content of the BX register. Assume that all numbers are unsigned integers.

The solution is given in listing 5.3 below.

```
            ;First - Set registers
    MOV DX,DIVIDEND_HIGH
    MOV AX,DIVIDEND_LOW
    MOV BX,DIVISOR
            ;Then perform division
    DIV BX
            ;and check for errors
```

Listing 5.3   16 bit division program.

Testing for errors following mathematical operations generally involves checking the overflow and carry flags. The division instruction also indicates a *divide by zero* error in a very dramatic way – by generating a special type of interrupt request!

Before moving-on to examine interrupt operation in detail, it is useful to mention the range of addressing modes available to the 8086 microprocessor. These are summarised in table 5.2.

Operand	Addressing mode
D8/D16	Immediate
M8/M16	Direct
R8/R16	Register direct
[ ]	Base or index (Register indirect)
[ ]M8/M16	Base or index plus displacement
[ ][ ]	Base plus index
[ ][ ]M8/M16	Base plus index plus displacement

Table 5.2   8086 addressing modes.

As can be seen, a range of complex addressing modes is available in which the effective address is formed by adding the content of up to two 16 bit registers, plus an optional 8/16 bit displacement.

Interrupt Types and Operation

The 8086 microprocessor supports three basic types of interrupt. These are

- non maskable interrupts,
- hardware generated maskable interrupts,
- software generated maskable interrupts.

A single non maskable interrupt is available, which is triggered by a low to high transition on the NMI pin. This generates a *type 2* interrupt, as explained below.

Hardware generated interrupt requests are produced by a high level on the INTR pin. The interrupt flag must be cleared in software to enable maskable interrupt requests, and this flag is automatically set during the ISR to disable further interrupt requests. The interrupt flag is restored to its previous state at the end of the ISR when the flags register is popped from the stack.

The 8086 uses a vectored interrupt system, and is capable of detecting up to 256 different interrupt types. Each interrupt vector is stored in memory as a 32 bit pointer. Two bytes define the code segment of the ISR, while the remaining 16 bits contain the instruction pointer value. Memory addresses 00000_{16} to $003FF_{16}$ hold the 8086 interrupt vector table, as shown in table 5.3

Number	Address	Description
0	00000	Divide by zero error
1	00004	Single step
2	00008	Non maskable interrupt
3	0000C	Breakpoint
4	00010	Arithmetic overflow
...	...	...
...	...	...
...		
	...	...
FF	003FC	...

Table 5.3 8086 interrupt vectors.

Following a successful interrupt request, the 8086 generates an *interrupt acknowledge cycle*. In response, the external device supplies a device identification byte which refers to one of the entries in the interrupt vector table. This number is multiplied by four to give the address of the appropriate entry in the interrupt vector table.

As mentioned previously, it is also possible to generate an interrupt in software, using an instruction of the form

```
INT  interrupt_number ;(0-255)
```

This facility is widely used as a mechanism for accessing operating system facilities in 'IBM compatible' microcomputers. For example, interrupts $5-1F_{16}$ are commonly used by the *BIOS* (*basic input/output system*) to perform low level, hardware-specific operations, while interrupt 21_{16} provides a large number of high level *DOS* (*disc operating system*) functions.

Particular DOS functions are accessed by loading the appropriate sub-function number into the AH register, before calling the interrupt. On return from the interrupt, the results (if any) are also passed via the data registers.

Example

DOS function 02H writes a single character to the standard output device, which in most cases is the screen. (It is also possible to redirect 'standard output' to devices such as printers and files.)

This function would be called using,

```
MOV AH,02H  ;DOS sub-function
INT 21H     ;Call DOS interrupt
```

(Those readers wishing to experiment with DOS or BIOS facilities for themselves may refer to any of the large number of DOS programmer's reference books which are available.)

Input/Output Port Addressing

The 8086 uses the lower 16 lines of the data bus, together with special control bus signals to provide 64 K of I/O port addresses (in addition to a maximum of 1 megabyte of physical memory).

Two different input instructions are available, each being quite different in operation. The

```
IN accumulator,port
```

instruction is used to input data from an immediate port address. The port address, which is specified as part of the instruction, is a single byte. This means that only the first 256 port addresses can be accessed using this instruction.

It is also possible to use register indirect addressing to specify the port address. In this case, the DX register holds a 16 bit port address, thus allowing the full I/O addressing capacity to be used. The normal use of this instruction is shown below.

```
MOV DX,port_address
IN accumulator,DX
```

A further two instructions are available for output instructions. These are

```
OUT port,accumulator
```

which sends data to an immediate port address in the range 0–255, and

```
OUT DX,accumulator
```

which uses the DX register to point to the port address.

Note here that (as was the case with memory considered earlier), even numbered port addresses are transferred via the lower half of the data bus, while odd addresses use the upper half of the data bus. Designers of hardware add-ons should be aware of the correspondence between odd and even port addresses and the physical connection to the data bus!

Case Study – the PIC 16C54

Compared with the rapidly expanding and increasingly complex Intel 80X86 family of microprocessors, the PIC family of *microcontrollers*, (which are produced by the American firm Microchip Technology Inc.) offers a complete contrast.

These extremely compact devices effectively provide a single IC computer solution which is ideally suited to *embedded control* applications. Typical uses of *PIC* (or *peripheral interface controller*) ICs include

- computer peripherals (keyboard decoder, digitiser and mouse),
- remote control transmitters (TV and video),
- domestic products (washing machine, microwave oven and security systems),
- consumer electronics (telephone, TV, video recorder, camcorder and intelligent toys),
- industrial electronics (sensors, data processing and PLCs),
- automotive electronics (electronic ignition, ABS, active suspension and car alarms).

A microcontroller contains all of the essential components of a microcomputer, housed within a single IC. This includes the microprocessor, data and program memory and input output ports. PIC microcontrollers in the 16C5X family also contain a real time clock/counter which may be used to measure accurately time intervals or count external pulses. Operation in electrically noisy environments, such as industrial or automotive applications, is also supported by the inclusion of a *watchdog timer*. If enabled, this may be used to reset the microcontroller automatically, should it 'hang-up' due to externally induced electrical noise.

Various types and sizes of program memory are available to suit different applications. EPROM memory is fitted to microcontrollers which are intended for prototype development, due its ability to be erased and later reused. Cheaper OTP (one time programmable) EPROMs are commonly used in mass produced designs. Table 5.4 shows the main features of the PIC 16C5X series.

Before examining the detailed internal operation of these devices, it is useful to understand the methods used to develop and test microcontroller based software.

A typical embedded system would not immediately be recognised as a microcomputer. Instead, the microcontroller is used to enhance a product's operation in some way. This may include a reduction in the product's size, chip count and cost. At the same time, the flexibility of the design is increased due to the ease with which the microcontroller can be reprogrammed.

Since the microcontroller is 'embedded' within the product, it is difficult to enter programs or examine the internal state of the device. For this reason, programs are normally written and tested on a different computer, known as a *development system*.

The assembly language *source code* is written using a text editor and is stored using the development system's filing system as an ASCII file. The source code is then assembled by a *cross assembler*, which produces a special type of *object code* file, containing the microcontroller machine code. (By definition, a cross assembler produces machine code that is not intended to run on the development system microprocessor.)

Testing of the software is normally performed on the development system using a *software simulator*, which mimics the microcontroller and its operating environment.

Having entered the simulator, the object code file produced by the cross assembler is loaded from backing store, after which it may be tested using suitable simulator commands. (Note that although software simulators are cheap and easy to use, the program cannot usually be executed at normal speed due to the complexity of the simulation task.)

Software that must be tested in *real time* may be transferred to an *in-circuit emulator* (*ICE*), which plugs into the microcontroller socket. This device emulates the external signals produced by the microcontroller, and can be controlled by the operator. A range of facilities is provided by the ICE which are similar to those of a monitor program. Facilities include register/memory display and edit, and program debug facilities such as single stepping or breakpoints. Programs may also be downloaded (or uploaded) from the development system using suitable commands.

Once the software has been tested, the microcontroller is then programmed using an EPROM programming unit. The correct operation of the microcontroller may then be confirmed in its intended environment, (which is commonly referred to as the *target system*).

The development system/target system approach to software generation and testing is commonly used where the target system lacks the facilities normally required for program development and testing. This approach is shown graphically in figure 5.10.

Device	Pins	I/O	RAM	EPROM	Frequency
PIC 16C54	18	12[†]	32 x 8	512 x 12	DC[‡]–20 Mhz
PIC 16C55	18	20[†]	32 x 8	512 x 12	DC[‡]–20 Mhz
PIC 16C56	18	12[†]	32 x 8	1K x 12	DC[‡]–20 Mhz
PIC 16C57	18	20[†]	80 x 8	2K x 12	DC[‡]–20 Mhz

Table 5.4   PIC 16C5X microcontroller family characteristics.

[†]   When configured to read externally applied pulses, the RTCC pin provides an additional input.

[‡]   Although the PIC is fully 'static', practical RC-based oscillator frequencies range from 25 kHz – 5 MHz, with higher frequencies requiring a quartz crystal controlled oscillator.

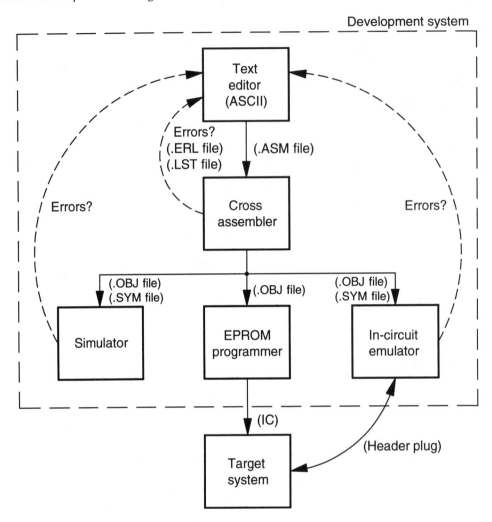

Fig. 5.10 Program writing and testing using a development system and target system approach.

As shown in figure 5.11, the PIC 16C54 has 12 input/output port lines, each of which may be individually configured as an input or output.

The two oscillator pins (OSC1 and OSC2) may be connected to an external RC network, or to a quartz crystal or ceramic resonator. RC networks are normally used in low frequency applications (25 kHz–5 MHz), where a wide frequency tolerance is acceptable, or where low power consumption is important. Reprogrammable (windowed) EPROM devices are configured for a particular oscillator type during programming, while OTP microcontrollers are normally tested for a particular oscillator type by the manufacturer.

The quartz crystal oscillator is normally used in high frequency applications, or where the frequency stability and tolerance of the oscillator are important.

In either case, the oscillator frequency is divided by four internally to produce four non-overlapping (*quadrature*) clock waveforms. The majority of PIC instructions are executed in a single machine cycle, which is equivalent to four clock cycles. Thus at 20 MHz, a single machine cycle lasts for 0.2 μs. Exceptions to this are jump instructions and bit test instructions (when true), which require two machine cycles.

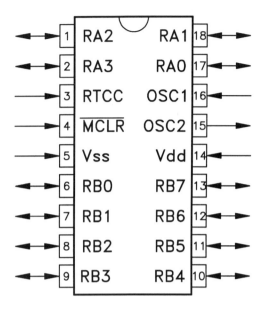

Fig. 5.11 PIC 16C54 pin connections.

Figure 5.12 gives typical RC oscillator connections and values, while figure 5.13 gives the same information for a crystal controlled oscillator.

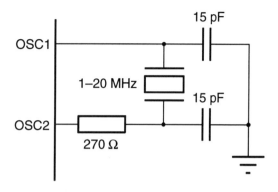

Fig. 5.13 Quartz oscillator circuit and values.

The $\overline{\text{MCLR}}$ pin provides the *power on reset* facility for the microcontroller. In cases where the rate of rise of V_{DD} is at least 0.05 volts/ms, $\overline{\text{MCLR}}$ may be directly connected to the positive supply. Following a logic 1 level on this pin, the microcontroller remains reset for a further 18 ms to allow the oscillator circuit to start up and stabilise. In cases where the oscillator is slow to stabilise or the power supply requirements are not met, the circuit of figure 5.14 may be used.

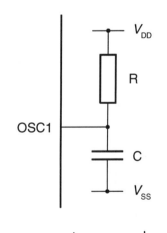

Frequency	R	C
5 MHz	3.3 kΩ	20 pF
1.2 MHz	5 kΩ	100 pF
75 kHz	100 kΩ	100 pF
28 kHz	100 kΩ	300 pF

Fig. 5.12 RC oscillator circuit and values.

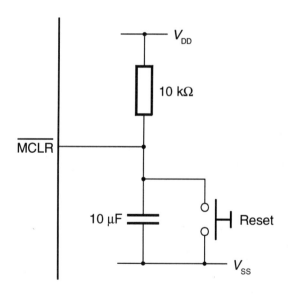

Fig. 5.14 Power on reset circuit with reset button (optional).

Lastly, a suitable d.c. power supply must be provided. This is nominally 5 volts, but some variation is allowed, depending on the oscillator type.

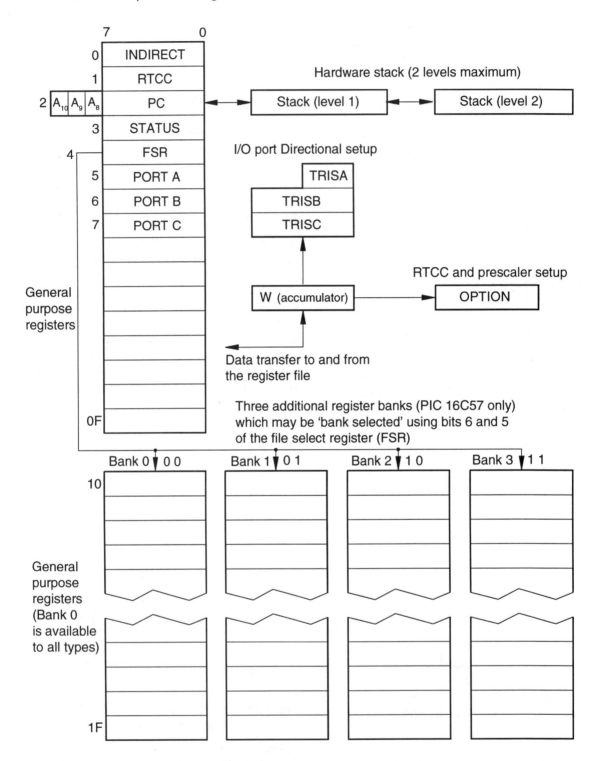

Fig. 5.15 PIC 16C5X internal register arrangement.

Figure 5.15 shows the internal register arrangement of the PIC 16C5X microcontroller family.

A bank of 32 registers is available to all microcontroller types, which is referred to as the *register file*. These are addressed as register locations $0–1F_{16}$. The 16C57 has an increased number of register locations, which is achieved by switching the lower half of the register file (locations $10–1F_{16}$), between four different register banks. At any time only one register bank is active, as selected by the state of bits 5 and 6 in the *file select register* (FSR).

Program instructions are stored in the EPROM memory, the size of which varies between different microcontrollers. With the PIC 16C54 and 16C55, the program memory consists of 512 locations, each of which is 12 bits wide. Simple arithmetic shows that the program counter must be 9 bits wide $(A_8–A_0)$ in this case, to allow each EPROM location to be addressed. As with the register file, the larger program memories of the PIC16C56 and 16C57 are accessed using a page swapping technique, which uses bits 5 and 6 of the status register to provide address lines A_9 and A_{10} respectively.

PIC microcontrollers actually use separate internal buses for transferring data and program words, which is known as a *Harvard architecture*. This contrasts with traditional or *von Neumann* microprocessors in which the same data bus is used to transfer op. codes and operands, but at different times. The advantage of using a Harvard architecture is that the instruction *pipelining* may be used to increase the operating speed of the microprocessor. As a consequence of the separate program and data buses, one instruction can be executed (using the data bus), while the next instruction is being fetched from program memory. The execute phase of one instruction overlaps with the fetch cycle of the next, making it appear that instructions are performed in a single machine cycle, while they actually require two!

The PIC 16C54 and 16C56 have 12 input/output port bits which are provided by ports A (4 bits) and B (8 bits). The PIC 16C55 and 16C57 also use port C (8 bits), giving 20 I/O bits in total.

Before a particular port is used for data transfer, the direction of each port bit must be initialised by writing to the appropriate I/O control register (TRISA, TRISB or TRISC). Writing a '1' to a particular bit, sets that bit as an input, while a '0' indicates an output.

When the ports are being initialised, the appropriate control word must first be loaded into the accumulator or W register. With port initialisation, this is normally performed using immediate addressing, as shown in listing 5.4.

```
                    ;Port names
Port_A equ 5
Port_B equ 6
Port_C equ 7
                    ;Port initialisation
    movlw 0fh       ;Move 0F (literal) to
                    ;the accumulator (W)
    tris Port_A     ;Configure port A
                    ;(all inputs)
    movlw 0         ;Move 0 (literal) to
                    ;the accumulator (W)
    tris Port_B     ;Configure port B
                    ;(all outputs)
    movlw 0ffh      ;Move FF (literal) to
                    ;the accumulator (W)
    tris Port_C     ;Configure port C
                    ;(all inputs)
```

Listing 5.4 Port initialisation routine.

An output port could then be written to by first loading an immediate value into the W register and then transferring the data to the appropriate port. This is illustrated by the following example.

```
    movlw 055h      ;Load data into W
    movwf Port_B    ;Transfer W to Port
                    ;B
```

Similarly, an input port could be read into W as shown below.

```
movf Port_B,0 ;Copy port B into
              ;the accumulator.
```

If the final argument is zero, the destination is the W register, while a value of 1 causes the value to be copied back into the original register (while updating the zero flag in the status register). This is normally written in a slightly more readable form as,

```
W equ 0          ;Assembler equate
  ..
  ..
  ..
movf Port_B,W ;Copy port B into
              ;the accumulator.
```

An individual bit of any register may also be set or cleared using the BSF (bit set) or BCF (bit clear) instructions. For example

```
bsf Port_B,3
```

may be used to set bit 3 of register 6 (port B). Similarly, the same bit could be cleared using

```
bcf Port_B,3
```

(The bit position must be a number in the range 0–7, with 0 being the least significant bit.)

Jump operations are performed using the GOTO instruction where the destination address is specified as a 9 bit operand. With the 16C56 and 16C57 microcontrollers, the most significant part of the destination address is specified by bits 5 and 6 of the status register.

The CALL instruction is used when a jump to a subroutine is made. Notice from figure 5.15 that (due to the lack of external RAM) PIC microcontrollers use a two level stack, which is internal to the microcontroller. This is commonly known as a *hardware stack*. A consequence of this arrangement is that only one level of subroutine nesting is permitted.

Another peculiarity associated with subroutines is that an 8 bit operand is used with the CALL instruction. This means that any subroutine used must have its entry address in the first 256 bytes of the selected program memory page!

A return from subroutine is performed using the RETLW instruction, which also has the facility to return a value to the calling program. For example the instruction

```
retlw 0fh
```

causes a return to the main program and places $0F_{16}$ into the W register.

Conditional jumps are also possible, although at first sight they do not appear to be part of the PIC instruction set. The BTFSC and BTFSS instructions are used to test the state of a particular bit of a register, and conditionally skip (or 'hop over') the next instruction. BTFSC skips the next instruction if the tested bit is clear, while BTFSS misses the next instruction if the tested bit is set.

An obvious application of this feature is to perform a conditional jump based on the state of a status register flag (whose internal arrangement is shown in figure 5.16 below).

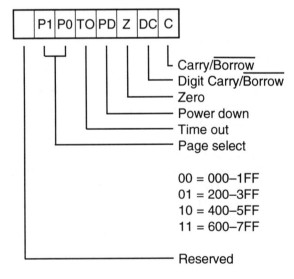

Fig. 5.16 PIC status register.

For example, a conditional jump could be performed, based on the state of the zero flag, as shown below.

```
                   ;Register/flag names
Zero equ 2
Status equ 3
...
...
                   ;Jump if zero
btfsc Status,Zero
goto set
...
...
                   ;Jump if not zero
btfss Status,Zero
goto clear
```

Notice that although bit testing of the status register has been shown here in association with jump instructions, this feature may be used to execute any PIC instruction conditionally, based on the state of any register bit!

A consequence of the program counter being part of the register file is that PIC instructions may operate directly on the PC. For example, the result of adding a literal value to register 2 (and also placing the result in the same register) is to cause a jump to a new address! This powerful feature makes it possible for the programmer to develop custom addressing modes by directly manipulating the program counter content.

It should also be noted here that operations which write to the program counter can only modify the lower 8 bits of the PC. The most significant bit of the PC is always cleared, forcing the destination address to be in the first 256 locations of the currently selected program memory page.

The PIC instruction set also supports indirect addressing using register 0, together with the lower 5 bits of the file select register. Any instruction that uses register 0 as an operand will actually specify an indirect address. For example, if the

This indirect approach is particularly useful where several registers must be accessed in sequence using a program loop. This is achieved by incrementing (or decrementing) the FSR after each operation.

Example

The following example demonstrates the direct modification of the program counter, based on a value that has been obtained by reading an input port. This causes a jump to one of 16 possible addresses, based on the 4 bit code present on port A.

As mentioned previously, the PIC 16C54 has 12 input/output port bits, as shown in figure 5.17

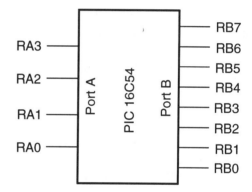

Fig. 5.17 PIC 16C54 input/output ports.

Assuming that port A is configured as an input and port B as an output, the device may be considered as a *combinational logic system* with four inputs and eight independent outputs.

Listing 5.5 demonstrates how the PIC 16C54 may be configured to operate as a 3 to 8 line decoder, with inputs RA2–RA0 providing the address select inputs and RA3 acting as an active low chip enable input.

```
        ;PIC register names
        include "picreg.equ"
        ;Macro definition
term    macro   input, output
        org     8 + (input * 10h)
        movlw   output
        movwf   Port_B
        goto    loop
        endm
        ;Start of main program
        org     PIC54
        goto    start
        org     0
start   movlw   0
        tris    Port_B
                ;RB0-RB7 outputs
        movlw   0fh
        tris    Port_A
                ;RA0-RA3 inputs
loop    movf    Port_A,w
                ;read port A
        swapf   Port_A,w
                ;swap nibbles in W
        andlw   0f0h
                ;mask upper nibble
        addwf   PC
                ;jump to output
        ;"Truth table" using the
        ;TERM macro definition
        term 00h,b'11111110'
        term 01h,b'11111101'
        term 02h,b'11111011'
        term 03h,b'11110111'
        term 04h,b'11101111'
        term 05h,b'11011111'
        term 06h,b'10111111'
        term 07h,b'01111111'
        term 08h,b'11111111'
        term 09h,b'11111111'
        term 0ah,b'11111111'
        term 0bh,b'11111111'
        term 0ch,b'11111111'
        term 0dh,b'11111111'
        term 0eh,b'11111111'
        term 0fh,b'11111111'
        end
```

Listing 5.5 Simulating combinational logic.

The program uses the *macro* facility of the PIC assembler to shorten the program and improve its readability. (Recall that *macro assemblers* were discussed in the first volume.)

Sixteen copies of the TERM macro are used, with one for each of the possible input combinations. The value obtained from port A is first multiplied by 16 (by swapping nibbles in the accumulator) before being added to the program counter. This direct modification of the PC produces a jump to the appropriate output routine.

Notice here that any 4 bit combinational logic function can be produced by this program, simply by editing the output codes defined by the macros. A useful technique is therefore to place the 'truth table' in a separate file and use an include statement to merge the two program sections during assembly. This approach is illustrated in listing 5.6.

```
        include "comb_log.h"
        term 00h,b'11111110'
        term 01h,b'11111101'
        term 02h,b'11111011'
        term 03h,b'11110111'
        term 04h,b'11101111'
        term 05h,b'11011111'
        term 06h,b'10111111'
        term 07h,b'01111111'
        term 08h,b'11111111'
        term 09h,b'11111111'
        term 0ah,b'11111111'
        term 0bh,b'11111111'
        term 0ch,b'11111111'
        term 0dh,b'11111111'
        term 0eh,b'11111111'
        term 0fh,b'11111111'
        end
```

Listing 5.6 Using a combinational logic header file.

Note that the above file would normally have a '.ASM' file extension for compatibility with the PIC assembler.

Practical Exercise	**Squaring Function**

Method

Write PIC software to simulate a combinational logic function where the 8 bit output produced is the 'square' of the 4 bit input.

For example, if the binary equivalent of 9 (1001_2) is applied to the inputs, the output produced should be 81_{10} (01010001_2).

Hint

Base your solution on the program of listing 5.6.

```
include "comb_log.h"
term 00h,b'00000000'
term 01h,b'00000001'
  . . .

  . . .
end
```

Example

Having considered the simulation of combinational logic, this example considers *sequencer* design, in which a series of codes are output from an output port in a repeating sequence.

Sequencer design is performed here by writing a table, where each line specifies the name of the current state, the name of the next state, and the code that should appear on the output terminals. Once again, port B is assumed to be an output, allowing the PIC to produce an 8 bit number at each step in the sequence. Listing 5.7 shows the required table for a repeating 3 bit Gray code sequence.

```
include "seq_log.h"
state s1,s2,b'00000000'
state s2,s3,b'00000001'
state s3,s4,b'00000011'
state s4,s5,b'00000010'
state s5,s6,b'00000110'
state s6,s7,b'00000111'
state s7,s8,b'00000101'
state s8,s1,b'00000100'
end
```

Listing 5.7 Gray code sequence generator.

Once again, the macro facility of the PIC assembler is used to simplify the design, with all PIC related information being placed in the header file of listing 5.8.

```
        ;PIC register names
            include "picreg.equ"

        ;Macro definition
state   macro    current,next,output
current movlw    output
        movwf    Port_B
        goto     next
        endm

        ;Start of main program
        org      PIC54
        goto     start

        org      0
start   movlw    0fh
        movlw    0
        tris     Port_B
        ;RB0-RB7 outputs
loop
```

Listing 5.8 Sequencer header file.

Practical Exercise   Electronic Dice

Method

Using sequencer design techniques, program a PIC 16C54 microcontroller to produce an electronic dice (1 or 2 digits).

Modify the sequencer header file of listing 5.8 to initialise port A as an input (in addition to port B being configured as an output).

The 'dice throw' function is implemented by making the following addition to the STATE macro definition.

```
movwf Port_B     ;old
btfss Port_A,1   ;new
goto loop        ;new
btfss Port_A,0   ;new
goto current     ;new
goto next        ;old
```

This gives:

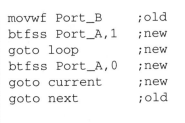

RA1	RA0	Function
0	X	Reset (display off)
1	1	Counting
1	0	Freeze display

The required state table (for 1 dice) is

```
include "seq_log.h"
state reset,s1,b'11111111'
state s1,s2,b'11111110'
state s2,s3,b'11110111'
state s3,s4,b'11111100'
state s4,s5,b'11110101'
state s5,s6,b'11110100'
state s6,s1,b'11110001'
end
```

Dice Spot Numbers

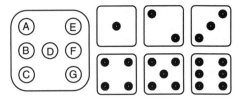

Output Circuit

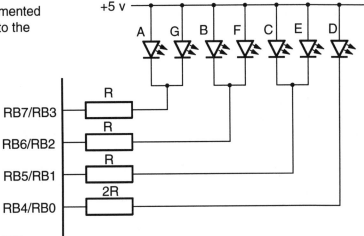

Throw Circuit

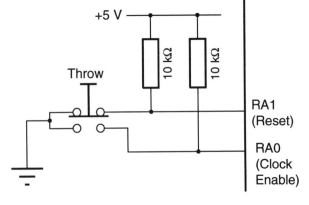

Note. $I_{OL(MAX)}$ = 25 mA, with a 50 mA limit for the entire port.

Two remaining features of PIC microcontrollers which have not yet been discussed are the *real time clock/counter* (RTCC) and the *watchdog timer*. As will be seen, these features are closely related.

The watchdog timer is intended for use in electrically noisy environments and causes the microcontroller to be automatically reset in the event of a program 'crash'. (Note that sources of electrical interference and methods of protecting circuits from its effects are considered in the next chapter).

Real time clock/counter support ICs were discussed in chapter 4 and it was seen that these devices are capable of counting externally applied pulses or accurately measuring time intervals.

The watchdog timer facility may be enabled or disabled during device programming by setting or clearing a special fuse, which is in a section of the EPROM not accessible during normal program operation. When enabled, the watchdog timer has a nominal time period of 18 ms, but this may be altered using an optional *prescaler*, giving a maximum time interval of approximately 2.5 seconds.

During normal operation, with the watchdog timer enabled, the program should issue the *clear watchdog timer* (CLRWDT) instruction at regular intervals, which prevents the microcontroller from being reset. In the event of a program crash, the watchdog timer continues to run, and once its time interval has elapsed, the microcontroller is automatically reset.

Another possible cause of a watchdog timer reset is the SLEEP command, which causes the microcontroller to enter a special operating mode with very low power consumption. All normal operation (including the oscillator) ceases once SLEEP mode has been entered, but the watchdog timer continues to run.

Following a device reset, it is possible to determine the cause of the reset by examining the TO (time out) and PD (power down) bits in the status register (which then allows the program to take appropriate action). The significance of these bits following a reset is given in table 5.5.

TO	PD	Meaning
0	0	WDT timeout during SLEEP
0	1	WDT timeout during 'normal' program operation
1	0	MCLR reset during SLEEP
1	1	Reset following power-up

Table 5.5 PD and TO significance following a device reset.

Prescaler settings are defined in the OPTION register, which is shown in figure 5.18.

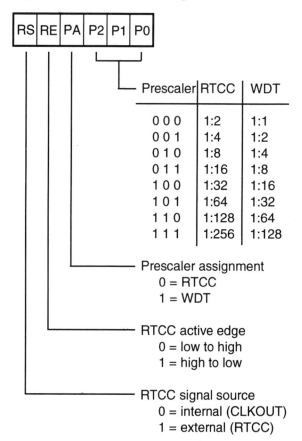

Fig. 5.18 OPTION register.

As can be seen, the prescaler can be assigned to either the watchdog timer, or to the RTCC, but not to both simultaneously. Once the prescaler assignment has been made, the least significant 3 bits define the division ratio used by the prescaler.

By default, following a reset, all bits of the OPTION register are set, thus selecting the watchdog timer with a 1:128 division ratio (producing a time period of 18 ms × 128 which equals 2.3 seconds, if enabled).

The RTCC feature may be selected by writing an appropriate control word to the OPTION register, as shown in listing 5.9.

```
movlw   07    ;Prescaler = 1:256
option        ;Count up on CLKOUT
```

Listing 5.9 Selecting the RTCC option.

In this case, the prescaler has been set to 1:256 using an internal signal source. The result is that register 1 (RTCC) is incremented after each 256 instruction cycles. Once the register reaches FF_{16} it changes to 0 on the next increment.

The content of the RTCC register may be read or altered at any time by the program, thus enabling time intervals to be accurately measured or external events to be counted. Note that unlike the RTCC ICs considered in the previous chapter, the PIC version does not have an interrupt facility. This means that a polling routine must be used to detect the required time interval or count.

Example

PIC microcontrollers are ideally suited to battery powered and portable applications, due to their small size and low power consumption. (Surface mount versions are also available, allowing very compact designs to be produced).

This example demonstrates the use of a PIC 16C54 as a hand-held infrared transmitter, which would be suitable for use in car alarm and remote control applications. The RTCC facility, derived from the system clock, is used to provide all timing information.

The solution given here incorporates several interesting features.

- Transmitter battery life is conserved by using very narrow transmit pulses.

- Information is represented by the time interval between pulses, rather than by a logic level. A short gap represents a logic 1, with a longer space indicating a logic 0.
- Due to the use of the previous pulse to provide a timing reference for the next pulse, the method used is relatively insensitive to differences in oscillator frequency between the transmitter and receiver. Cumulative timing errors, such as those found in asynchronous serial data transfer cannot occur, which allows cheaper timing components to be used.
- The receiver must receive the same code twice, otherwise a data transmission error is assumed to have occurred. This feature can virtually eliminate data errors, since the code length is determined by the programmer.

Table 5.6 shows the RTCC timing values used by the transmitter and receiver.

Type	Transmitter	Receiver
Pulse	1	1
Mark (logic 1)	4	2–7
Space (logic 0)	16	8–31
Pause	64	32–127

Table 5.6 Infrared transmitter/receiver timing.

Figure 5.19 shows the appearance of a 4 bit data word, encoded using this method.

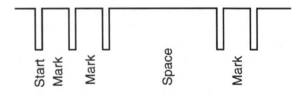

Fig. 5.19 4 bit data word (1101).

The above code would be repeated, following a 'pause', to verify correct data transmission, and would be initiated by resetting the microcontroller using a switch. Power would then be conserved by entering SLEEP mode, immediately following the data transfer.

The transmitter circuit used here is based on an infrared LED, as shown in figure 5.20.

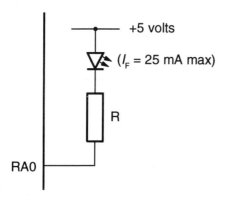

Fig. 5.20 Infrared transmitter circuit.

Once again, the macro facility of the PIC assembler has been used to simplify the writing of the transmitter program. Listing 5.10 gives an example transmitter program for the code '101100', while listing 5.11 shows the necessary header file.

```
        include "transmit.h"
        start        ;Begin first sequence
        mark
        space
        mark
        mark
        space
        space

        pause        ;Short delay

        start        ;Begin second sequence
        mark
        space
        mark
        mark
        space
        space

        sleep
        end
```

Listing 5.10 Example transmitter program.

```
        include  "picreg.equ"
start   macro                   ;Macro def's
        local    loop
        clrf     RTCC
        movlw    0
        movwf    Port_B ;Start pulse
loop    movlw    1
        subwf    RTCC,W
        btfss    STATUS,Z
        goto     loop
        movlw    1
        movwf    Port_B ;End pulse
        endm
mark    macro
        local    loop
        clrf     RTCC
loop    movlw    4
        subwf    RTCC,W
        btfss    STATUS,Z
        goto     loop
        start
        endm
space   macro
        local    loop
        clrf     RTCC
loop    movlw    10h
        subwf    RTCC,W
        btfss    STATUS,Z
        goto     loop
        start
        endm
pause   macro
        local    loop
        clrf     RTCC
loop    movlw    40h
        subwf    RTCC,W
        btfss    STATUS,Z
        goto     loop
        endm
        org      PIC54   ;Main prog.
        goto     begin
        org      0
begin   movlw    0
        tris     Port_B ;B = output
        movlw    3
        option           ;Prescaler
```

Listing 5.11 Transmitter header file.

Practical Exercise   Infrared Alarm/Immobiliser

Method

Using a PIC 16C54 microcontroller, design a car alarm/immobiliser that is enabled or disabled by the remote control transmitter discussed in the main text.

When armed, the alarm should produce a running light display on the 8 LED outputs as a visual deterrent. The immobiliser relay should also be active at this time.

Block Diagram

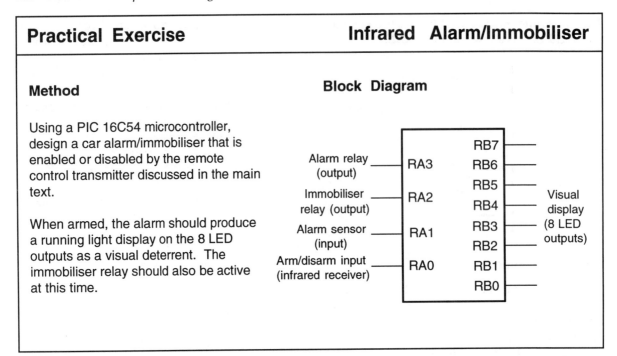

Alarm relay (output) — RA3
Immobiliser relay (output) — RA2
Alarm sensor (input) — RA1
Arm/disarm input (infrared receiver) — RA0

RB7
RB6
RB5
RB4
RB3
RB2
RB1
RB0

Visual display (8 LED outputs)

Simulating Program Operation

Once assembled, the program's operation may be *simulated* before proceeding to program the PIC's internal EPROM, using the supplied programming unit. The ability to simulate the program's operation is very useful, given the difficulty of examining the internal state of the actual system. (As mentioned earlier, simulation is a facility often provided by development systems.)

Once the simulator has been entered, the object code produced by the assembler may be loaded and tested using suitable simulator commands. These include the ability to display the content of registers and ports, as well as program testing facilities such as single stepping and breakpoints. External signal transitions may also be injected into the simulator from a *stimulus file*, which allows the program's response to external events to be tested.

Listing 5.12 shows a typical *initialisation file* which may be used to automate much of the simulator configuration.

```
p 54
ad trisa,b,4
ad trisb,b
gs Port_A,5,f
ad Port_A,b,4
gs Port_B,6,f
ad Port_B,b
ad W
sc 0.2
lo list6
verbose on
st list13
ss
```

Listing 5.12 Typical simulator initialisation file.

The initialisation file first defines the processor type (16C54), before enabling display variables TRISA, TRISB, Port_A and Port_B. (Additional variables may be added, if required.) The processor instruction cycle time is then set to 0.2 µs (20 MHz clock), before the appropriate object code and stimulus files are loaded. Lastly, single step mode is entered, using the *SS* command.

The content of the stimulus file should be carefully chosen to allow the operation of the program to be fully tested. As an example, listing 5.13 shows a stimulus file which could be used to test 'combinational logic' designs, such as the 3 to 8 line decoder given in listing 5.6.

STEP	RA3	RA2	RA1	RA0
0	0	0	0	0
20	0	0	0	1
40	0	0	1	0
60	0	0	1	1
80	0	1	0	0
100	0	1	0	1
120	0	1	1	0
140	0	1	1	1
160	1	0	0	0
180	1	0	0	1
200	1	0	1	0
220	1	0	1	1
240	1	1	0	0
260	1	1	0	1
280	1	1	1	0
300	1	1	1	1

Listing 5.13 Example stimulus file.

This section has introduced the major features of the PIC 16C5X series of microcontrollers, which allow compact and high performance microelectronic systems to be developed at relatively low cost. For reference purposes, tables 5.7, 5.8 and 5.9 list the 33 instruction types which make up the PIC 16C5X instruction set.

Instruction	Function Performed
ADDWF f,d	Add W and f
ANDLW k	AND literal and W
ANDWF f,d	AND W and f
CLRF f	Clear f
CLRW	Clear W
CLRWDT	Clear Watchdog timer
COMF f,d	Complement f
DECF f,d	Decrement f
DECFSZ f,d	Decrement f, skip if zero
INCF f,d	Increment f
INCFSZ f,d	Increment f, skip if zero
IORLW k	Inclusive OR literal and W

Instruction	Function Performed
IORWF f,d	Inclusive OR W and f
MOVLW k	Move literal to W
MOVF f,d	Move f
MOVWF f	Move W to f
NOP	No operation
RLF f,d	Rotate f left, through carry
RRF f,d	Rotate f right, through carry
SUBWF f,d	Subtract W from f
SWAPF f,d	Swap nibbles in f
XORLW k	Exclusive OR literal and W
XORWF f,d	Exclusive OR W and f

Table 5.7 Byte-oriented operations.

Instruction	Function Performed
BCF f,b	Clear bit b in f
BSF f,b	Set bit b in f
BTFSC f,b	Test bit b in f, skip if clear
BTFSS f,b	Test bit b in f, skip if set

Table 5.8 Bit-oriented operations.

Instruction	Function Performed
CALL k	Call subroutine
GOTO k	Go to new address (9 bit)
OPTION	Load option register from W
RETLW k	Return form subroutine, placing literal value in W
SLEEP	Enter standby mode
TRIS f	Configure port directions (f = 5, 6 or 7)

Table 5.9 Control operations.

Notes

k = literal value (8/9 bit)
f = register number ($0–1F_{16}$)
b = bit number (0–7)
d = destination (0 = W, 1 = source register)

Summary

The main points covered in this chapter are:

- Owing to increases in program size and complexity, the address bus width of modern microprocessors has steadily increased. The Intel 80486 for example has a 32 bit address bus allowing 4 GB to be directly addressed. Larger program sizes can also be accommodated using *virtual memory* techniques in which only the active section of a large program needs to be in memory at any time.
- Data bus widths have increased steadily, as this allows the data throughput of the microprocessor to be increased for a given clock speed. Intel's Pentium microprocessors for example have a 64 bit data bus.
- Increasing microprocessor clock speeds have allowed more instructions to be executed in a given time, thus increasing the processing speed of the microprocessor. Intel 80486DX/4 microprocessors for example operate (internally) at up to 100 MHz.
- *RISC* design techniques have optimised the efficiency of modern microprocessors, resulting in faster operation and lower heat dissipation (due to the simpler design).
- The latest microprocessors (including Intel Pentium devices) have multiple *execution units* and use *superscalar* techniques to execute more than one instruction simultaneously from the *instruction queue*.
- Some microprocessors (such as the Intel 8085 and 8086) use a time-multiplexed address and data bus to reduce the required number of external connections.
- *Harvard architecture* microprocessors use separate program and data buses, allowing the internal operation of the microprocessor to be optimised. *Instruction pipelining* allows one instruction to be fetched while another is simultaneously executed.
- *Microcontrollers* are being used increasingly in *embedded control* applications due to their small size and low cost.

Problems

1   Explain current and future trends in microprocessor design, related to bus widths and speed of execution of programs.

2   Define the following terms.

   a) Virtual memory
   b) Multitasking
   c) Multiplexed data bus
   d) RISC microprocessor
   e) Superscalar microprocessor
   f) Harvard architecture

3   Explain the following terms, as related to the Intel 8086 microprocessor.

   a) Execution unit
   b) Bus interface unit
   c) Instruction queue
   d) Segment register

4   Define the main features of a microcontroller and give reasons why these devices are being used increasingly in embedded control applications.

6   Principles of System Design

Aims

When you have completed this chapter you should be able to:

1 Relate schematic circuit diagrams to printed circuit board layouts.
2 Appreciate the available types of PCB and basic manufacturing techniques.
3 Understand the effects of resistance, capacitance, inductance, electrical noise and transmission line effects on practical circuit performance.
4 Use good design techniques to minimise these problems.

Introduction

This chapter begins by considering the types of diagram that are commonly used to represent electronic circuits and the use of *printed circuit board (PCB)* techniques during their manufacture.

The often unwanted electrical effects that may arise from poor circuit design and PCB layout are then considered. These include track resistance, capacitive and inductive coupling between tracks, susceptibility to noise and signal reflections in high speed circuitry.

Methods of minimising these effects by good design are then outlined. Techniques considered include the use of decoupling capacitors, line drivers, terminated lines, PCB layout strategies, shielding and the use of watchdog timers in electrically noisy environments.

Schematic and Logic Diagrams

Schematic diagrams are a commonly used method of representing electronic circuit designs. A typical example is shown in figure 6.1 below.

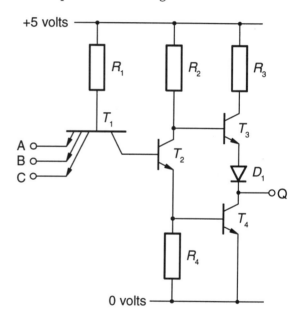

Fig. 6.1 A typical schematic diagram.

Notice firstly that all components are sequentially numbered, allowing them to be referred to in any related text. For example, a *parts list* could be used to specify relevant parameters for each item, including component values, tolerances, part number, cost etc. Explanations of circuit function or fault-finding schedules may also refer to these same component numbers.

Circuit nodes or connections between components are shown using a 'dot', which normally joins three wires. To avoid ambiguity, connections between four wires are normally avoided, as shown below.

3 wire connections

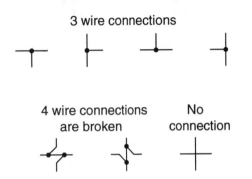

4 wire connections are broken   **No connection**

Fig. 6.2   Representing schematic connections.

A variation on the schematic diagram, often encountered in digital electronics is the *logic diagram*. Figure 6.3 shows a typical example.

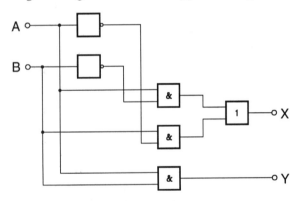

Fig. 6.3   Typical logic diagram.

Logic diagrams are used to represent the internal arrangement of digital circuitry (a half adder circuit in this example).

Details such as power supply connections are normally omitted, as it is the circuit's function that is emphasised, rather than its detailed construction. It should also be noted that several gates would be combined within a single IC in a practical circuit. (In fact, the circuit might be converted to a NAND-only or NOR-only solution, or even designed using a custom logic IC to minimise production costs!)

Constructional details may easily be added to the previous diagram, as shown below.

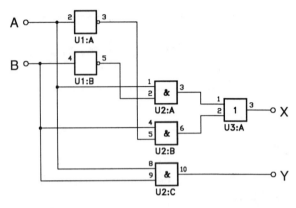

Fig. 6.4   The addition of pin and gate numbers.

Printed Circuit Board Methods

Although schematic and logic diagrams are useful methods of representing electrical and electronic circuits, they cannot be used directly in the manufacture of a circuit. Before proceeding to the manufacturing stage, several important issues must be resolved including the

- physical dimensions of each component (including pin numbers and positions),
- chosen constructional technique (prototype board, stripboard or printed circuit board for example),
- method used to connect all inputs, outputs and power supplies to the circuit board (plugs, connectors, wiring harnesses etc.),
- relative position or layout of each component on the circuit board (hence determining the minimum size of the board),
- actual size of the circuit board, plus the position and size of any fixing holes (which may also relate to the box or housing chosen for the circuit).

Clearly, a great deal of thought must be given to these questions, in order to avoid problems later!

The majority of electronic components have pin spacings which, for historical reasons, are multiples of 0.1 inches, as shown in figure 6.5.

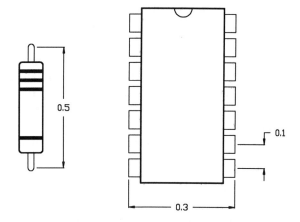

Fig. 6.5 Typical 0.1 inch component pin spacings.

This allows the circuit designer to plan the placement of components using a 0.1 inch grid, as shown below. (A smaller grid must be used for *surface mount* components, due to their 0.05 inch or 0.025 inch pin spacings.)

When deciding on the relative position of each component, a useful rule is to place interconnected components as close as possible to each other. This tends to minimise the length of any tracks or links between components and also reduces electrical interference, as will be seen later. In many cases, the schematic diagram will naturally have been laid out in this way, thus providing a good starting point.

If the ideal layout is not obvious, a *connection matrix* may be formed to count the number of connections between each component. A suitable layout may then be chosen, ensuring that those components with the largest number of connections are placed in the closest possible proximity. Table 6.1 shows an example matrix.

	IC1	IC2	IC3	Con. 1
IC1	-	2	2	4
IC2		-	1	0
IC3			-	3
Con. 1				-

Table 6.1 Typical connection matrix.

Integrated
circuit   Resistor   Capacitor

× ×
× × ● × × ● × × × ● × × × ● × × × × × × ×
× × ● × × ● × × × × × × × × × × × × × × ×
× × ● × × ● × × × × × × × ● × × × × × × ×
× × ● × × ● × × × × × × × × × × × × × × ×
× × ● × × ● × × × × × × × × × × × × × ● ×
× × ● × × ● × × × ● × × × × × × × × ● × ×
× × ● × × ● × × × × × × × × × × × × × ● ×
× ×

Transistor

Fig. 6.6 Planning component layout on a 0.1 inch grid.

In the previous example, the largest number of connections is seen to be between connector 1 and ICs 1 and 3. This suggests that connector 1, IC1 and IC3 should be placed as close together as possible. A possible arrangement is shown in figure 6.7.

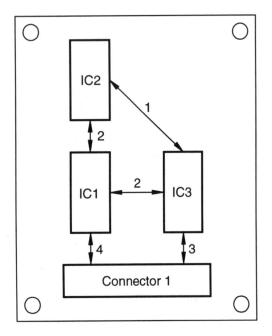

Fig. 6.7 Ideal component placement.

Once the component placements have been decided, it becomes relatively straightforward to select the overall size of board, and the placement of any mounting holes.

When deciding on the method of manufacture, the ideal solution will depend on the answer to the following questions.

● Has the design been proven? If not, then use a plug-in prototype board (or *circuit simulation software*) to test the circuit, before proceeding to final production.
● How many circuits will be manufactured? For one or two circuits, a stripboard design might represent the best option. For larger quantities, PCB design becomes much more cost effective, due to the possibility of mass production.

Several basic types of printed circuit board are available to the designer. The method chosen will depend on the complexity of the design, available PCB manufacturing facilities, and any cost/performance design constraints.

Single Sided PCBs

This is the simplest and cheapest type of circuit board. The PCB itself is made from an insulating *laminate* (normally a glass fibre and epoxy resin mixture), which is coated with a thin layer of copper on one side.

Components are placed on one side of the board, with the leads pushed through pre-drilled holes in the laminate. Unwanted areas of copper are removed using a chemical etching process, leaving the desired circuit connections or *tracks*. Component leads are soldered in place to copper *pads*, which are placed at the ends of each track. These soldered joints perform the dual function of providing a low impedance electrical connection between the component and the track, and also hold the component in place. (Heavy components may need additional fixing to prevent damage to the PCB.) Figure 6.8 shows a cross-sectional view of a single sided PCB.

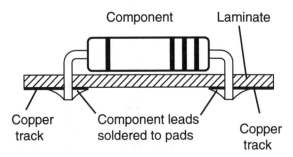

Fig. 6.8 Single sided PCB.

Initially, the entire track side of the PCB is covered in a thin layer of copper. PCB production involves the production of an etch resistant coating on those areas of the PCB which are to be retained, either using a photographic technique or by drawing directly on the surface with a special pen. Uncoated areas are then stripped away using an etchant solution.

The use of an *etch resist pen* to draw the required track pattern directly onto the PCB is only suitable for one-off production, and the final result tends to be of poor quality. Once the track layout has dried, the PCB is normally immersed in a solution of ferric chloride (for approximately 20 minutes) to remove unwanted areas of copper. After etching, the board is drilled, components are inserted, and then soldered into place.

Where more than one PCB is to be manufactured, and quality is important, a photographic production technique is normally used. In this case the PCB is coated with a light sensitive emulsion, which is exposed to ultraviolet light inside a sealed *light exposure unit*. A transparent acetate sheet, containing the desired track layout, is placed over the photo-sensitive PCB during exposure. The acetate sheet contains a positive image of the PCB, allowing unwanted areas to be 'developed' by exposure to the ultraviolet light.

The photo sensitive coating in these developed areas is then stripped away by immersing the PCB in a dilute alkaline solution, such as sodium hydroxide. At this point, the copper side of the PCB is covered by a positive photographic image. The board may then be etched using ferric chloride and then assembled, as explained above. Further boards may then be manufactured by reusing the acetate film.

Several methods exist for the production of the desired track layout on acetate film.

- *Rub-on transfers* to produce the desired pads and tracks. This may be performed at a 1:1 scale or, in the case of intricate designs, may be designed at 2:1 scale and then photographically reduced to actual size.
- *Computer aided design* (*CAD*) software to draw the PCB layout. Low cost (manual design) systems are available, in which the user is responsible for all design decisions. More expensive (auto-routing) software allows the process of component placement and track routing to be automated, to a greater or lesser extent.

In this case, the circuit design is initially entered in schematic form. The PCB design software then generates a *netlist* (a list of circuit connections), which is used by the autorouter section of the program. (This netlist may also be passed to *circuit simulation* software, which may be used to verify correct operation of prototype designs before final PCB production.)

Output from such a CAD system may be printed onto high temperature acetate sheet using a laser printer, or could be directly drawn using a pen plotter.

Professional PCB manufacture may involve several additional stages, including the production of solder resist patterns, drilling details and silk screen masks.

In mass production, the PCB is soldered in a single operation by passing it over a bath of molten solder, rather than by hand soldering individual components. This is known as the *flow solder* technique. Solder is prevented from sticking to unintended areas by a (normally green) *solder resist* coating, which covers all parts of the PCB, except the pads.

A *drilling detail* is used to specify the x,y coordinate and diameter of each hole on the PCB, which is a requirement for professional PCB production. In most cases, a numerically-controlled drilling machine is used for this purpose. Some types of PCB design software automatically produce drilling details, or even a *part program*, which may be used to directly control the machine tool.

A *silk screen mask* is normally printed onto the component side of the PCB, which indicates the names, positions and correct orientations of components. This is useful for hand assembled PCBs, although it is less vital where robotic component insertion techniques are used.

Double Sided PCBs

With a complex PCB design, it may be impossible to route all tracks using a single sided PCB. Particularly difficult tracks may need to be completed by the addition of wire links or *jumpers* soldered to the component side of the PCB. In this case a double sided PCB may represent a better solution.

As its name suggests, a double sided PCB has tracks on both the component and track sides of the board, as shown in figure 6.9.

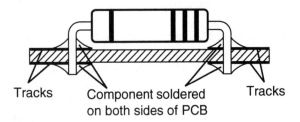

Fig. 6.9 Double sided PCB.

Production of a double sided PCB begins with a copper clad laminate which has a photo sensitive coating on both sides. The PCB is then exposed to ultraviolet light in a double sided light exposure unit. Great care must be taken at this stage with the alignment of the two acetate films, to ensure correct registration of tracks and pads on each side of the board.

In amateur and semi-professional production, component leads are normally soldered on both sides of the PCB. This provides electrical continuity between tracks on the component and track sides of the board. In cases where a track has to swap to the other side of the board, in the absence of a component lead (commonly called a *via*), a small *PCB pin* may be inserted and soldered in place, as shown in figure 6.10.

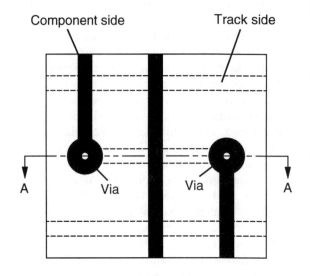

Sectional view though A-A

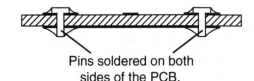

Pins soldered on both sides of the PCB.

Fig. 6.10 Use of PCB pins in double sided boards.

In professionally manufactured double sided PCBs, an electrical plating process is used to deposit a layer of copper inside each via. These *plated-through* holes provide a direct electrical connection between tracks on the component and track sides of the PCB, thus eliminating the need to solder component leads on both sides of the board, or to use PCB pins. The first step during PCB manufacture is to drill the required via holes, which are then electrically plated. The board is then coated with a light sensitive emulsion, after which the board is developed and etched. (A range of professional PCB manufacturing processes exist which differ in detail, though not in principle, from the previously discussed manufacturing techniques.)

Each plated-through hole adds a small amount to the cost of manufacturing the PCB, so the number of vias is normally minimised. One useful technique is to place tracks which are predominantly horizontal on one side of the board, with mostly vertical tracks on the other side. More difficult connections which cross the PCB diagonally may then be resolved into horizontal and vertical components. Each track section is placed on the appropriate side of the board, and is linked by a via. This is the basic technique used by many auto-routing PCB design programs.

Multiple Layer PCBs

Although the double sided PCB with through-plated holes is a great improvement on the single sided PCB, the increasing component packing densities of modern designs frequently require the use of multiple layer boards.

Depending on the complexity of the design, a four, six, or even ten layer PCB may be needed. With a four layer PCB, such as that shown in figure 6.11, two layers would be devoted to power supply connections, with the remaining two layers used for signal tracks.

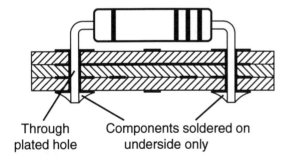

Through plated hole   Components soldered on underside only

Fig. 6.11 Four layer, through plated PCB.

The PCB itself is formed by bonding together several thin laminates, with through-plated holes forming the connection between tracks on different layers. These complex PCBs offer a considerable performance advantage over single or double sided designs, but are more expensive.

Flexible PCBs

Flexible printed circuit boards are used in many electronic products, particularly where space is at a premium, or cost is an important factor. In many cases, the PCB is used directly to form connections to various points, as well as being used to hold any electronic components. The elimination of individual connectors, combined with an inherently low profile allows cost effective and compact designs to be produced, which are particularly suited to the consumer electronics industry. Flexible PCBs are available in single sided, double sided, or multiple layer forms, and using a variety of base materials.

Designers should ensure that boards are not permitted to bend in areas where components or soldered pads are placed, due to the risk of mechanical failure. If necessary, stiffening materials may be fixed to the board in these areas.

Electrical Parameters of PCBs

When components are interconnected according to a circuit diagram, the physical arrangement of components and their connections will inevitably lead to the creation of parasitic resistors, capacitors or inductors. Even though these components do not actually appear on the circuit diagram, the designer should be aware of their existence and of ways of minimising their effects.. To ignore these unwanted effects may result in improper or unreliable circuit operation.

For example, a long thin copper track on a PCB may posses significant resistance, which in turn may result in an unintended volt-drop along its length.

An unwanted capacitance is created by two parallel tracks, where each track acts as one of the capacitor plates. A sudden change in voltage on one track may cause interference on the other, due to capacitive coupling between the tracks.

A current carrying wire (particularly a current carrying loop) is also found to posses a finite inductance. Changing the current flowing through the loop leads to a changing magnetic flux. This changing flux may link with other current carrying loops in the circuit, leading to interference due to inductive coupling.

Although not normally mounted on the PCB itself, it must also be realised that the d.c. power supply normally possesses a large self inductance. For this reason, a sudden change in the current demanded by the circuit may lead to large supply line transients.

In addition to the creation of unwanted components on the PCB itself, the very high operating frequencies of modern electronic circuits may lead to further problems. This is because the inductance and capacitance associated with each track cause the connection to act as a *transmission line*, which will have an associated *characteristic impedance*. If the *terminating impedance* of the track differs from the characteristic impedance, then unwanted *signal reflections* will occur. This effect is visible by examining the rising and falling edges of digital signals, and is commonly referred to as *ringing*.

The following sections describe each of these effects in detail and suggest ways of minimising their impact.

Resistance of a Track

An important property of a track is the *resistivity* of the conducting material from which it is made. Resistivity may be defined as the electrical resistance presented by a one metre cube of the material. The resistivity of copper, for example, is typically given as 0.0173×10^{-6} Ωm.

The actual resistance of a track is based not only on the material, but also on its length and cross-sectional area. It is found experimentally that increasing the cross-sectional area of a conductor reduces its resistance, while increasing the length has the effect of increasing the resistance. This is expressed mathematically by equation 6.1.

$$\text{Resistance} = \frac{\text{Resistivity} \times \text{Length}}{\text{Area}}$$

or
$$R = \frac{\rho \, l}{A} \qquad (6.1)$$

Example

Calculate the resistance of a 0.5 mm wide copper track, which is 250 mm long. Assume the average depth of copper on the PCB is 0.035 mm.

The resistance is given by,

$$R = \frac{0.0173 \times 10^{-6} \times 0.25}{0.035 \times 0.5 \times 10^{-6}}$$

$$= 0.25 \; \Omega$$

Although a resistance of 0.25 Ω may seem insignificant, it may cause an unwanted volt drop along the track, particularly if the current flowing through the track is large. For this reason, the width of power supply and ground tracks is normally made much wider than that of low current signals. Track resistance may also be minimised by placing components as close together as possible.

Capacitive Coupling Between Tracks

The capacitance of a parallel plate capacitor is given by,

$$C = \frac{\varepsilon_0 \, \varepsilon_r \, A}{D} \qquad (6.2)$$

where,

ε_0 = Permittivity of free space (8.85×10^{-12} F/m).
ε_r = Relative permittivity of the dielectric.
A = area of the plates (m^2).
D = distance between the plates (m).

Now consider the capacitance between two tracks on opposite sides of the PCB, as shown in figure 6.12.

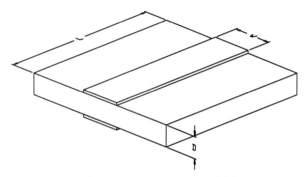

Fig. 6.12 Capacitance between PCB tracks.

The capacitance of this arrangement may easily be calculated. It is assumed that the area (A) of the plates is given by the product of the length and width which is common to both tracks. The distance between the plates (D) is given by the thickness of the PCB, while the relative permittivity of the PCB depends on the chemical composition of the laminate. (The value of ε_r is likely to range between 4.2 to 5.3 for epoxy resin/glass fibre PCBs.)

Example

Calculate the capacitance per cm between two tracks, each 1 mm wide and separated by a distance of 1.5 mm. Assume that the PCB laminate has a relative permittivity of 5.0.

$$C = \frac{8.85 \times 10^{-12} \times 5 \times 1 \times 10^{-3} \times 1 \times 10^{-2}}{1.5 \times 10^{-3}}$$

$$= 0.3 \text{ pF/cm} \qquad (1 \text{ pF} = 10^{-12} \text{ farads})$$

Assuming that capacitive coupling exists between two tracks, a sudden change in the potential of one track may cause a voltage spike on the other. This is because the charge stored by the parasitic capacitor cannot instantly be dissipated. Since $Q = CV$, a change in the potential of one plate of the capacitor results in a similar change on the other plate!

Although this example has considered the capacitance between tracks on opposite sides of a PCB, it should be realised that capacitive coupling also exists between adjacent tracks on the same side of the board. In general, capacitive coupling between tracks may be reduced by

- minimising track length while maximising the distance between tracks,
- avoiding the use of excessively wide signal tracks, which tends to increase the track area, and hence its capacitance,
- deliberately providing capacitive coupling to ground, either by the use of a *ground plane* (a large grounded area, often on the opposite side of the PCB), or by placing a grounded track between the two signal connections on the same side of the board.

Inductive Coupling Between Tracks

Any flow of electric current results in the generation of a magnetic field around the conductor, as shown in figure 6.12.

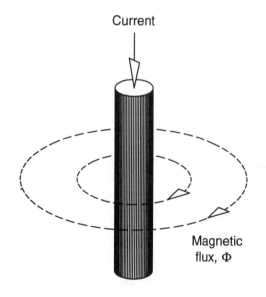

Current

Magnetic flux, Φ

Fig. 6.13 Magnetic field around a conductor.

In practice, it is difficult to imagine an isolated piece of wire carrying a current, since electrical current normally flows around a closed circuit, or loop. Figure 6.14 illustrates the magnetic field produced by a single current-carrying loop of wire.

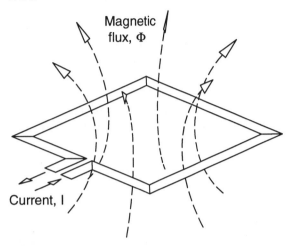

Fig. 6.14 Magnetic field of a single loop of wire.

Clearly, if the current changes, then so will the magnetic flux. It is found experimentally that an e.m.f. is induced in a conductor whenever it 'cuts' lines of magnetic flux. Furthermore, the direction of this e.m.f. acts to circulate a current which opposes the original change in magnetic flux. This self-induced e.m.f. is related to the change in current through the conductor by the *self inductance* (L), as shown in equation 6.3.

Self induced e.m.f. $E = L \dfrac{di}{dt}$ (6.3)

If the changing flux caused by one loop cuts another conductor, or other conducting loop, an e.m.f. will be induced in the second circuit. The magnitude of this e.m.f. is found to be proportional to the rate of change of the magnetic flux which links the two circuits. The two circuits are said to possess *mutual inductance*. The e.m.f. induced in the second circuit is given by

Mutually induced e.m.f. $E_2 = L_{12} \dfrac{di_1}{dt}$ (6.4)

This inductive coupling between tracks works in both directions. A changing current in circuit 1 will induce an e.m.f. in circuit 2, and vice versa. Thus a circuit which generates electromagnetic interference is also prone to receiving it!

In practice, the self inductance of a loop is proportional to its area, and mutual inductance between current carrying loops is proportional to the 'shared' area between loops.

One simple way to reduce the area of a loop is to place the sending and return conductors of the loop as closely together as possible. One way to think of this is that, because of the opposite directions of current in the two wires, the magnetic field produced by one wire tends to cancel that produced by the other. The small loop area also makes the loop relatively immune to externally generated magnetic fields.

For example, the inductance of the power supply loop may be minimised by placing the supply and ground tracks as closely together as possible. On a double sided PCB, this may be achieved by placing each track on opposite sides of the board, as shown in figure 6.15.

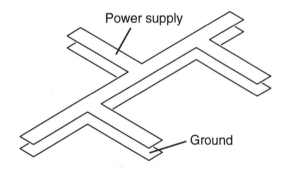

Fig. 6.15 Good placement of power supply and ground tracks.

In practice, it may be difficult to identify harmful loops which may produce or receive magnetic interference. A useful technique is to place a large conducting area or ground plane on one side of the PCB, which acts as a return path to ground. Returning currents will tend to flow directly underneath the supply track since this minimises the loop area and is the lowest energy solution.

Power Supply Inductance

The current demanded by digital ICs rises momentarily as internal transistors change state. This is partly due to the need to charge and discharge parasitic capacitances inside each IC, and is related to the switching speed of the device. Logic families with high power consumption and fast switching speeds tend to cause very severe current transients on the power supply lines.

As seen in the previous section, a sudden change in the current consumption of an IC may cause a change in the power supply voltage, due to the self inductance of PCB tracks. (A typical PCB track has an inductance of about 10 nH/cm). These momentary supply fluctuations may result in unreliable circuit operation, which may be difficult to trace. In addition, small current transients produced by individual ICs, may result in much larger changes in the current consumed by the PCB itself. These much larger current transients must be supplied directly by the power supply circuitry, which has a large self inductance. Clearly a voltage transient on the main power supply rail is very serious, due to its direct connection to all parts of the circuit!

As seen in the previous section, the inductance of tracks may be reduced by minimising the area of any current carrying loops, but this is only a partial solution.

A second technique is to place *decoupling capacitors* at strategic points on the PCB. Each capacitor is connected across the power supply rails and acts as a reservoir of charge. Short-lived current transients are supplied by the decoupling capacitor, rather than by the power supply, effectively eliminating the harmful effects of power supply transients.

Placement of decoupling capacitors is very important. A single large capacitor is generally placed close to the point where d.c. power is supplied to the PCB. The position of this capacitor is not critical, with a 10–100 μF electrolytic capacitor being typical. This capacitor acts to recharge the much smaller decoupling capacitors, which are placed at various points on the board.

Small decoupling capacitors are placed in close proximity to ICs, in order to supply locally generated current transients. Typically, one decoupling capacitor is placed for every couple of ICs, with 0.1 μF ceramic capacitors being commonly used. (Electrolytic capacitors are not suitable in this application, due to their rather high inductance.) Each capacitor must be placed as close as possible to the IC, in order to minimise the inductance of the connections between the capacitor and the IC. Figure 6.16 illustrates a good power supply and decoupling capacitor arrangement.

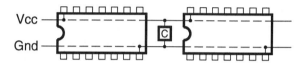

Fig. 6.16 Good decoupling capacitor layout.

The lead inductance associated with the decoupling capacitor effectively forms a series LC circuit, whose resonant frequency is given by,

$$F_0 = \frac{1}{2\pi\sqrt{(LC)}} \tag{6.5}$$

At frequencies below resonance, the circuit is primarily capacitive and functions as a decoupler. However, at higher frequencies the circuit becomes primarily inductive and ceases to perform any useful function. The product of LC must be minimised, which may be achieved by reducing the size of L, and/or of C.

Example

Calculate the resonant frequency of a decoupling circuit whose capacitance and lead inductance are 0.1 μF and 20 nH respectively.

$$F_0 = \frac{1}{2 \times 3.142 \times (2 \times 10^{-9} \times 0.1 \times 10^{-6})^{0.5}}$$

$$= 11.25 \text{ MHz}$$

(Decreasing C by a factor of 100 would increase the resonant frequency by ten times.)

Transmission Line Effects

At very high operating frequencies the electrical properties of PCB connections, such as inductance and capacitance, become important. When operating at these speeds, the PCB becomes an active component, whose electrical characteristics must be considered. In fact, the inductance and capacitance which is distributed along the length of a track, will cause a pair of parallel tracks to act as a *loss-free transmission line* (assuming the series resistance and conductance can be ignored).

A basic transmission line property is the *characteristic (or iterative) impedance*. This is the impedance seen 'looking into' the input terminals of the transmission line, when the line is terminated by an impedance which is also equal to the characteristic impedance. This value may be determined mathematically, or by trial and error (which explains the alternative term). A typical transmission line is illustrated by figure 6.17.

Inductance and capacitance
per unit length

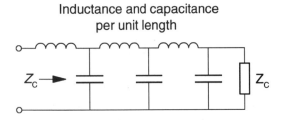

Fig. 6.17 Loss-free transmission line.

For a loss-free transmission line, the characteristic impedance is given by,

$$Z_c = \sqrt{\frac{L}{C}} \qquad (6.6)$$

Signals travel down the transmission line at a speed which is a fraction of the speed of light (c). In high speed circuits, this delay between a signal being sent and received must be taken into account. In addition, it is found that if the transmission line is terminated by an impedance other than the characteristic impedance, then *signal reflections* may occur. Any *discontinuity* in the transmission line may cause the signal to be partially or wholly reflected.

A signal may be reflected several times and 'bounce' back and forth along the transmission line, growing smaller and smaller. This transient effect is known as *ringing* and is a feature of an incorrectly terminated transmission line. Figure 6.18 illustrates the effect of ringing on an ideal pulse waveform.

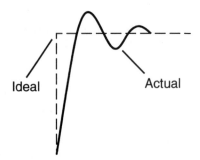

Fig. 6.18 Ringing due to signal reflection.

As a general rule, if the time taken for the signal to propagate down the transmission line is greater than one third of the signal rise or signal fall time, then transmission line effects should be taken into account. The signal velocity is given by,

$$u = \frac{c}{\sqrt{(\mu_r \, \varepsilon_r)}} \quad \text{(m/s)} \qquad (6.7)$$

where μ_r and ε_r are the relative permeability and permittivity of the transmission line and c is the velocity of light in a vacuum (3×10^8 m/s).

Example

Calculate the time taken for a signal to travel along a transmission line of length 150 mm, assuming that $\mu_r = 1$ and $\varepsilon_r = 5$.

$$u = \frac{3 \times 10^8}{\sqrt{5}} = 1.34 \times 10^8 \text{ m/s}$$

Since time = distance/velocity...

$$t = \frac{0.15}{1.34 \times 10^8} = 1.1 \text{ ns}$$

The exact nature of the reflected signal depends on the value of the load impedance. Assuming the forward travelling voltage and current are called V^+ and I^+, and the load impedance is resistive, the reflected voltage and current (V^- and I^-) will be given by,

$$V^- = \frac{R - Z_c}{R + Z_c} V+ \qquad (6.8)$$

$$I^- = -\frac{R - Z_c}{R + Z_c} I+ \qquad (6.9)$$

Assuming values of L and C of 8 nH/cm and 0.4 pF/cm respectively gives a characteristic impedance of 140 Ω (from equation 6.6), although practical values for Z_c may vary widely from this.

The load impedance of the transmission line may in practice be the input impedance of an integrated circuit, at the other end of the PCB connection. If this input impedance is greater than the characteristic impedance then the reflected voltage will be of the same polarity as the forward-travelling pulse (while the current has the opposite polarity). This may cause the receiving-end voltage to be momentarily greater than or less than the supply voltage.

Several methods of preventing unwanted signal reflections, or ringing, are in common use. The use of *buffer* or *line driver* ICs, as shown in figure 6.19 is a commonly used technique.

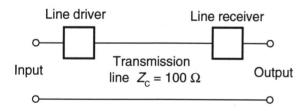

Fig. 6.19 Using line driver ICs to eliminate ringing.

In this case, the output and input impedances of the line driver/receiver ICs are matched to that of the transmission line, to avoid ringing.

Other methods include the modification of the receiving-end input impedance, by the use of additional components, as shown in figure 6.20.

General TTL method

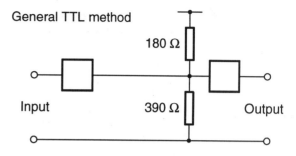

AC termination (TTL or CMOS)

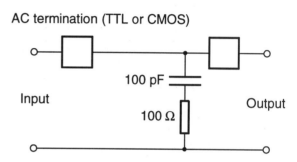

Fig. 6.20 Bus termination techniques.

Another method is to place a resistor in series with the output, at the sending end. This normally has a value which is 10 Ω less than the characteristic impedance of the line, and matches the output impedance of the gate to the transmission line.

A disadvantage of these methods is the increased power supply drain, combined with the difficulty of determining the characteristic impedance in each case. An alternative is to use *clamping diodes* at the receiving end. These are normally connected across the power supply and prevent the receiving end voltage from becoming significantly greater than or less than the supply voltage limits. Arrays of clamping diodes are also effective at eliminating crosstalk and electrical noise. Figure 6.21 illustrates a 16 bit Schottky barrier diode bus termination array (reproduced by courtesy of Texas Instruments).

SN74S1053
16-BIT SCHOTTKY BARRIER DIODE BUS-TERMINATION ARRAYS

D3424, SEPTEMBER 1990

- **Designed to Reduce Reflection Noise**

- **Repetitive Peak Forward
  Current . . . 200 mA**

- **16-Bit Array Structure Suited for
  Bus-Oriented Systems**

- **Package Options Include Plastic "Small
  Outline" Packages and Standard Plastic
  300-mil DIPs**

DW OR N PACKAGE
(TOP VIEW)

V_CC	1	20	V_CC
D01	2	19	D16
D02	3	18	D15
D03	4	17	D14
D04	5	16	D13
D05	6	15	D12
D06	7	14	D11
D07	8	13	D10
D08	9	12	D09
GND	10	11	GND

description

These Schottky barrier diode bus-termination
arrays are designed to reduce reflection noise on
memory bus lines. These devices consist of a
16-bit high-speed Schottky diode array suitable for
clamping to V_{CC} and/or GND.

The 74S1053 is characterized for operation from
0°C to 70°C.

schematic diagram

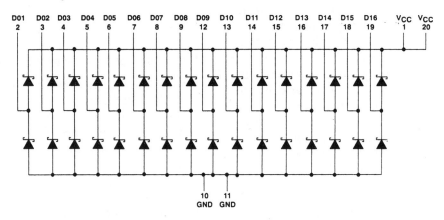

**TEXAS
INSTRUMENTS**

Fig. 6.21 (a) 16 bit Schottky barrier diode bus termination arrays (courtesy of Texas Instruments).

SN74S1053
16-BIT SCHOTTKY BARRIER DIODE BUS-TERMINATION ARRAYS

D3424, SEPTEMBER 1990

absolute maximum ratings over operating free-air temperature range (unless otherwise noted)†

Steady-state reverse voltage, V_R	7 V
Continuous forward current, I_F: any D terminal from GND or to V_{CC}	50 mA
total through all GND or V_{CC} terminals	170 mA
Repetitive peak forward current‡, I_{FRM}: any D terminal from GND or to V_{CC}	200 mA
total through all GND or V_{CC} terminals	1 A
Continuous total power dissipation at (or below) 25°C free-air temperature (see Note 1)	625 mW
Operating free-air temperature range	0°C to 70°C
Storage temperature range	−65°C to 150°C

† Stresses beyond those listed under "absolute maximum ratings" may cause permanent damage to the device. These are stress ratings only and functional operation of the device at these or any other conditions beyond those indicated under "recommended operating conditions" is not implied. Exposure to absolute-maximum-rated conditions for extended periods may affect device reliability.
‡ These values apply for $t_w \leq 100$ μs, duty cycle ≤ 20%.
NOTE1: For operation above 25°C free-air temperature, derate linearly at the rate of 5 mW/°C.

electrical characteristics over recommended operating free-air temperature range (unless otherwise noted)

single-diode operation (see Note 2)

PARAMETER		TEST CONDITIONS		MIN	TYP§	MAX	UNIT
V_F	Static forward voltage	To V_{CC}	$I_F = 18$ mA		0.85	1.05	V
			$I_F = 50$ mA		1.05	1.3	
		From GND	$I_F = 18$ mA		0.75	0.95	V
			$I_F = 50$ mA		0.95	1.2	
V_{FM}	Peak forward voltage		$I_F = 200$ mA		1.45		V
I_R	Static reverse current	To V_{CC}	$V_R = 7$ V			5	μA
		From GND				5	
C_T	Total capacitance	$V_R = 0$,	$f = 1$ MHz		8	16	pF
		$V_R = 2$ V,	$f = 1$ MHz		4	8	

§ All typical values are at $V_{CC} = 5$ V, $T_A = 25$°C.
NOTE 2: Test conditions and limits apply separately to each of the diodes. The diodes not under test are open circuited during the measurement of these characteristics.

multiple-diode operation

PARAMETER		TEST CONDITIONS		MIN	TYP§	MAX	UNIT
I_X	Internal crosstalk current	Total $I_F = 1$ A,	See Note 3		0.8	2	mA
		Total $I_F = 198$ mA,	See Note 3		0.02	0.2	

§ All typical values are at $V_{CC} = 5$ V, $T_A = 25$°C.
NOTE 3: I_X is measured under the following conditions with one diode static, and all others switching: Switching diodes: $t_w = 100$ μs, duty cycle = 0.2; static diode: $V_R = 5$ V. The static diode's input current is the internal crosstalk current I_X.

switching characteristics over recommended operating free-air temperature range (unless otherwise noted) (see Figures 1 and 2)

PARAMETER		TEST CONDITIONS	MIN	TYP	MAX	UNIT
t_{rr}	Reverse recovery time	$I_F = 10$ mA, $I_{RM(REC)} = 10$ mA, $I_{R(REC)} = 1$ mA, $R_L = 100$ Ω		8	16	ns

TEXAS INSTRUMENTS

Fig. 6.21 (b)   16 bit Schottky barrier diode bus termination arrays (courtesy of Texas Instruments).

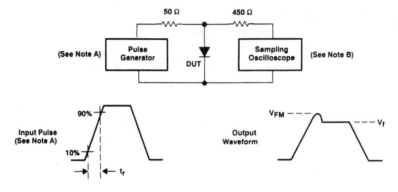

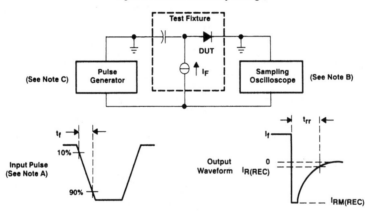

NOTES: A. The input pulse is supplied by a pulse generator having the following characteristics: t_r = 20 ns, Z_{out} = 50 Ω, f = 500 Hz, duty cycle = 1%.
B. The output waveform is monitored by an oscilloscope having the following characteristics: $t_r \leq$ 350 ps, R_{in} = 50 Ω, C_{in} = $\leq$ 5 pF.
C. The input pulse is supplied by a pulse generator having the following characteristics: t_f = 0.5 ns, Z_{out} = 50 Ω, t_w = $\geq$ 50 ns, duty cycle = 1 %.

TEXAS
INSTRUMENTS

3-859

Fig. 6.21 (c)   16 bit Schottky barrier diode bus termination arrays (courtesy of Texas Instruments).

SN74S1053
16-BIT SCHOTTKY BARRIER DIODE BUS-TERMINATION ARRAYS

D3424, SEPTEMBER 1990

TYPICAL APPLICATION INFORMATION

Large transients occurring at the inputs of memory devices (DRAMs, SRAMs, EPROMs, etc), or on the CLOCK lines of many clocked devices can result in improper operation of the devices. The SN74S1051 and SN74S1053 diode termination arrays help suppress transients caused by transmission line reflections, crosstalk, and switching noise.

Diode terminations have several advantages when compared to resistor termination schemes. Split resistor or Thevenin equivalent termination can cause a substantial increase in power consumption. The use of a single resistor to Ground to terminate a line usually results in degradation of the output high level, resulting in reduced noise immunity. Series damping resistors placed on the outputs of the driver will reduce transients, but they can also increase propagation delays down the line, as a series resistor reduces the output drive capability of the driving device. Diode terminations have none of these drawbacks.

The operation of the diode arrays in reducing transients is explained in the following figures. The diode conducts current whenever the voltage reaches a negative value large enough for the diode to turn on. Suppression of transients is tracked by the current-voltage characteristic curve for that diode. Typical current-voltage curves for the SN74S1051 / S1053 are shown in Figures 3 and 4.

To illustrate how the diode arrays act to reduce transients at the end of a transmission line, the test setup in Figure 5 was evaluated. The resulting waveforms with and without the diode are shown in Figure 6.

The maximum effectiveness of the diode arrays in suppressing transients occurs when they are placed at the end of a line and/or the end of a long stub branching off a main transmission line. The diodes can also be used to reduce the transients that occur due to discontinuities in the middle of a line. An example of this is a slot in a backplane that is provided for an add-on card.

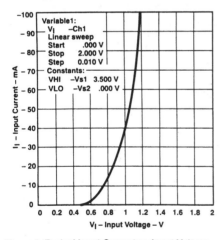

Figure 3. Typical Input Current vs Input Voltage
(Lower Diode)

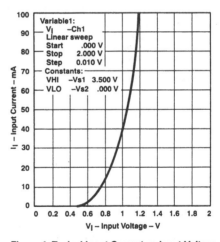

Figure 4. Typical Input Current vs Input Voltage
(Upper Diode)

TEXAS
INSTRUMENTS

Fig. 6.21 (d)   16 bit Schottky barrier diode bus termination arrays (courtesy of Texas Instruments).

SN74S1053
16-BIT SCHOTTKY BARRIER DIODE BUS-TERMINATION ARRAYS

D3424, SEPTEMBER 1990

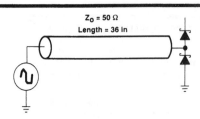

Figure 5. Diode Test Setup

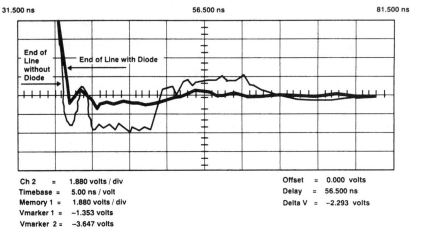

```
Ch 2      =   1.880 volts / div        Offset   =   0.000  volts
Timebase  =   5.00 ns / volt           Delay    =   56.500 ns
Memory 1  =   1.880 volts / div        Delta V  =   -2.293 volts
Vmarker 1 =   -1.353 volts
Vmarker 2 =   -3.647 volts
```

Figure 6. Scope Display

Fig. 6.21 (e)   16 bit Schottky barrier diode bus termination arrays (courtesy of Texas Instruments).

Sources of Electrical Noise

Digital circuitry is often thought of as being immune to the effects of electrical noise, but this is not always the case. In fact, a particular design may work perfectly in an electrically 'quiet' laboratory, but fail randomly when placed in an industrial or automotive environment. The design of reliable digital circuitry therefore depends on an understanding of sources of electrical noise and the methods of protecting against its effects!

It was seen in chapter one that TTL ICs have a *noise immunity* of 400 mV, which means that an output voltage will be correctly interpreted by an input, providing the noise voltage is less than 0.4 volts. A transient and unpredictable noise signal may easily exceed this limit, leading to unreliable circuit operation.

Noise voltages can sometimes be large enough to do physical damage to electronic circuitry, in addition to the creation of data errors. For example, a person walking across a nylon carpet may generate a static charge of 40 kV or more. An *electrostatic discharge (ESD)* will then occur when that person touches an object, such as a computer keyboard. This may easily cause static damage to input circuitry, or cause the computer program to 'crash'. An obvious defence against ESD is the use of clamping diodes, on circuits that may be susceptible to static discharge.

As mentioned previously, the inductance associated with mains power supplies may cause large e.m.f.s to be induced when industrial or office equipment is switched on or off. Transient voltages of 1 kV or more may be found on the mains supply in these circumstances, which in turn is fed to all mains-powered equipment. This is an example of *conducted noise*, meaning that the noise arrives at the victim circuit through electrical wiring. The defence in this case is the use of filters or suppressors on the power supply input circuitry.

Noise may also appear in the from of electromagnetic radiation, which is known as *radiated noise*. E.m.f.s are induced in the victim circuit which leads to the circuit failure.

Sparks or arcs may be generated by a range of electrical and electronic devices including switches, automotive ignition circuitry, motor commutators arc welding equipment and lightning. These may lead to the production of *electromagnetic pulses (EMP)* or *radio frequency interference (RFI)*, which may in turn be picked up by digital circuitry.

The cure in this case is to use *shielding* techniques, either to eliminate the radiation at source, or to prevent it from reaching the victim circuit. Radiated noise may consist primarily of magnetic fields, electric fields, or a mixture of the two.

A simple shield against electric fields is to enclose the circuit in a (usually earthed) metal box, commonly known as a *Faraday shield*. This provides capacitive coupling to ground, thus preventing the penetration of unwanted electric fields.

Magnetic screening is based on the minimisation of conducting loop areas at both sending and receiving ends, which reduces the mutual inductance between the two circuits. The use of a ground plane on the PCB is an effective defence in this case.

Electromagnetic screening uses similar techniques to the Faraday Shield mentioned above, although the electrical properties of the material used, and the method of construction (placement of joints and holes etc.) become more important.

In electromagnetic screening, an externally applied electromagnetic waveform causes currents to flow in the surface of the shield. This weakens the electromagnetic waveform, due to the action of *absorbtion* and *reflection losses*. *Absorbtion loss* (also known as I^2R *loss*) occurs due to the resistance of the shield material itself. Circulating currents in the shield also generate their own electromagnetic fields, which are then radiated from the surface of the screen. This second effect is known as *reflection loss*, for obvious reasons. Both effects are powered by the external waveform, thus causing the electromagnetic radiation to be weakened as it passes through the shield.

Poor earthing layout may also lead to noise problems. The earth circuit in poorly designed systems may contain mains 'hum' or high frequency current transients, which may upset normal circuit operation.

The reduction of cabling costs may suggest the series connection of power supply and ground connections to individual circuit boards, but this not always wise. Series connection means that large earth currents will flow through each board. Current transients from noisy circuits (such as multiplexed LED displays or other transistor switching circuitry) may then induce e.m.f.s in the victim circuit. This is another form of conducted noise, which may easily be avoided by the *star* connection of power supply and ground wiring.

Unwanted currents may also flow through the earth-ground if the circuit is physically connected to earth at two different points. This is due to the differing potentials at various points on the Earth's surface. Avoiding multiple connections to Earth is the solution in this case.

Noise Tolerant Computers

Inevitably, when computers are used in electrically noisy environments, there is a significant risk of failure due to transient noise. Normal operation may be resumed by pressing the reset button, although this is clearly inconvenient, and in some cases dangerous.

The process of resetting the computer following a crash may be automated by using a *watchdog timer* circuit. In this case, the computer program is often configured to produce regular pulses during normal operation. These are fed to a re-triggerable monostable (or missing pulse detector circuit), whose output remains set as long as the pulses continue. In the event of a crash, the pulses stop, causing the computer to be briefly reset by the watchdog timer.

Other useful techniques include the filling of unused memory locations with jump or restart instructions, which point to the main program. (Notice here that 00_{16} represents the 6502 BRK instruction, while FF_{16} is interpreted as RST 038H by the Z80!)

Summary

The main points covered in this chapter are:

- *Schematic diagrams* are often used to describe electrical and electronic circuits or design ideas. A variation on this theme is the *logic diagram*, which is used with digital circuits.
- The most commonly used manufacturing technique used for electronic circuitry is the *printed circuit board* or *PCB*, due to its suitability for mass production.
- A range of PCB types is available to the designer including single sided, double sided (with or without through-plated holes), multiple layer and flexible boards. The best approach for a particular design will depend on several factors including cost, overall dimensions and required component packing density.
- When planning track routing on double sided or multiple layer boards, a useful technique is to place horizontal connections on one layer and vertical tracks on another. Difficult connections may then be resolved into vertical and horizontal sections with a *through plated hole* or *via* used to link each section.
- Although PCBs are often thought of simply as a method of interconnecting components, the PCB itself possesses a range of electrical properties. These include track resistance, capacitive and inductive coupling between tracks and transmission line effects. The PCB designer should be aware of these effects, and wherever possible try to minimise them.
- An effective technique in all cases is to plan the layout so as to minimise the length of any connections between components. Drawing up a *connection matrix* may prove useful when deciding upon the ideal component layout.
- Power supply and ground connections should be made wider than signal tracks, to allow for their increased current-carrying capacity.

- Inductance associated with the power supply and ground tracks may be minimised by placing them as close together as possible (which minimises the power supply loop area). A possible solution with double sided boards is to place each supply track on opposite sides of the PCB. The creation of a large earthed area, or *ground plane* on one layer of the PCB is also effective.

- *Decoupling* capacitors may be used to eliminate transient current pulses due to IC switching. One large capacitor is normally connected where the power supply enters the PCB, with smaller capacitors placed close to small group of ICs. Close connection is important to minimise track inductance between the capacitor and each IC.

- Capacitive coupling can be minimised by maximising the distance between tracks. Excessive track widths should be avoided on signal connections, as this increases the area of the track and hence its capacitance. (Capacitance associated with wide power supply tracks is actually an advantage, since it provides power supply decoupling.)

- At high speeds, the capacitance and inductance associated with PCB tracks causes each signal connection to act as a *transmission line*. A typical PCB-based transmission line may have a characteristic impedance in the region of 100 Ω Signals may be partially reflected by any change in the impedance, such as an incorrect terminating load resistance. A typical symptom is the appearance of transient voltage waveforms or *ringing* on signals edges. Possible solutions include the use of *buffer* or *line driver* ICs, load impedance modification or the use of *clamping diode* arrays.

- There are several sources of electrical noise, each of which may cause circuit failure. These include *electrostatic discharge (ESD)*, *electromagnetic pulse (EMP)*, *radio frequency interference (RFI)* and earth-based noise. In general, *shielding* techniques are used against radiated noise, *filters* protect against conducted noise and good earth layout prevents transient earth currents.

- Several methods exist for protecting computer circuitry from the effects of electrical noise (in addition to shielding and good circuit design). One commonly used technique is the *watchdog timer*, which causes the computer to be reset if it fails to produce the expected series of output pulses.

Problems

1   Describe the main steps involved in the production of a single sided PCB.

2   Name five different types of PCB and suggest a suitable application for each type.

3   The following connection matrix has been produced from a schematic diagram.

	IC1	IC2	IC3	IC4	Con. 1
IC1	-	5	2	1	7
IC2		-	3	0	6
IC3			-	5	1
IC4				-	0
Con. 1					

Suggest a suitable PCB layout, based on the above connection matrix.

4   Calculate the resistance of a copper PCB track which is 0.25 mm wide and 200 mm long. Assume the average depth of copper is 0.035 mm and the resistivity of copper is 0.0173×10^{-6} Ωm.

5   a) Calculate the capacitance between the power supply and ground tracks, which have been placed on opposite sides of a PCB. The PCB thickness is 1.5 mm and each track is 200 mm long and 5 mm wide.
    b) Is capacitive coupling between power supply tracks a disadvantage? (Explain your answer.)

6   What steps would you follow, when designing a PCB layout, in order to minimise unwanted circuit effects such as track resistance and inductive and capacitive coupling between tracks?

7   An electrostatic discharge (ESD) is caused by a person touching a bare wire after building up a static charge. A current pulse is produced which has a rise time of 4 A/ns.

    a) If the wire has an inductance of 10 nH/cm, calculate the volt-drop in each cm of wire.
    b) Suggest a method of protecting computer input circuitry from the effects of ESD.

8   Explain how a design may be protected from induced e.m.f.s, which may be caused by sudden changes in the current demanded by the PCB (due to switching of digital ICs for example).

9   a) Calculate the characteristic impedance of a PCB connection, assuming the capacitance and inductance are 0.5 pF/cm and 6 nH/cm respectively.
    b) Explain how the production of unwanted signal reflections or ringing may be avoided in a practical circuit design.

10 a) Name five possible sources of electrical noise, which might cause a digital electronic circuit to malfunction.
    b) Suggest methods of defending your circuit from each of these noise sources.

7   Practical Applications

Aims

When you have completed this chapter you should be able to:

1 Appreciate the advantages of using multiplexing with multiple-digit 7-segment displays.
2 Input and output analogue signals using analogue to digital and digital to analogue converter ICs.
3 Control the speed, direction and position of d.c. and stepper motors using open loop and closed loop techniques.
4 Use phase and integral cycle control methods to vary the power supplied to an a.c. load.

Display Multiplexing

Several methods of interfacing 7-segment displays to a microcomputer were discussed in the first volume. At that time, the most efficient technique considered was to use a binary to 7-segment decoder IC which allowed two 7-segment displays to be driven from an 8 bit output port.

Using this method, a four digit display, for example, would require 16 output port bits, which is rather wasteful of computer resources. There is also duplication of components, as each display requires its own resistor array and decoder IC. The multiplexing technique considered here offers a considerable saving in the hardware required, for a small increase in software complexity.

Multiplexed displays use switching techniques to illuminate each 7-segment display in a repeating sequence. If this process is performed at sufficient speed, the human eye considers the display to be continuously illuminated.

The *persistence* of human vision means that if the display is updated fifty times per second, then no display flicker will be evident to the user. (Other examples of this principle are television pictures which are built up using a repeating scanning process, and lighting which is driven from the a.c. mains supply.) Figure 7.1 shows the scanning sequence for a 4 digit display.

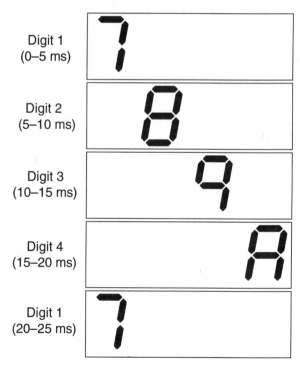

Digit 1 (0–5 ms)

Digit 2 (5–10 ms)

Digit 3 (10–15 ms)

Digit 4 (15–20 ms)

Digit 1 (20–25 ms)

Fig. 7.1 Display multiplexing cycle.

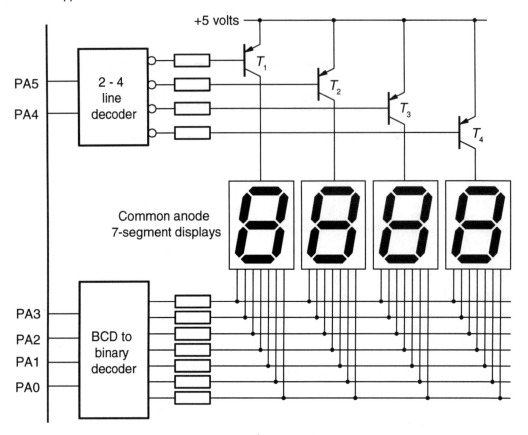

Fig. 7.2 Multiplexed 7-segment display circuit.

The actual circuit required for a multiplexed 7-segment display is shown in figure 7.2. Notice that only six output port bits are required to drive this circuit, compared to the 16 required by a non-multiplexed display. There are also savings in the number of ICs and resistor arrays required.

A 2 bit code is applied to the 2 to 4 line decoder, which causes the selected decoder output to become logic 0. This means that only one of the transistors (T_1–T_4) can be switched on at any time. By applying an incrementing count to bits PA4 and PA5, each transistor may be enabled in a repeating sequence.

When a particular transistor is enabled, it allows current to flow into the common anode terminal of the selected 7-segment display. The character displayed then depends on the 4 bit code applied to bits PA0–PA3, which in turn drives the decoder IC.

A possible display algorithm is given below.

1 Send first BCD code to PA3–PA0 and 00_2 to PA5–PA4, then wait for 5 ms.
2 Send second BCD code to PA3–PA0 and 01_2 to PA5–PA4, then wait for 5 ms.
3 Send third BCD code to PA3–PA0 and 10_2 to PA5–PA4, then wait for 5 ms.
4 Send fourth BCD code to PA3–PA0 and 11_2 to PA5–PA4, then wait for 5 ms.
5 Goto step 1 always.

The above algorithm is ideally suited for implementation as an interrupt service routine, possibly triggered by an interval timer IC. Communication between the main program and the ISR would then occur by writing to a shared area of memory, which would be used as a *display buffer* (either 2 or 4 bytes depending on whether data is stored in packed or unpacked BCD form).

Analogue Signal Interfacing

Until now, input or output of information has been exclusively digital, with all values restricted to either logic 0 or logic 1. When communicating with the 'real world', where many quantities are analogue in nature, it is often essential to be able to read and generate analogue quantities.

There are many possible types of analogue or digital input or output, each of which may be interfaced to a computer using suitable circuitry. Some of the more common signal types are given in table 7.1, together with typical input *sensors* and output *actuators* used in each case.

(The terms 'sensor' and 'transducer' are used interchangeably here.)

In many cases, a particular operating principle may be adapted to provide either analogue or digital information, depending on the application. For example, fluid depth could be measured using one or more float-activated switches, thus providing digital information only. Alternatively a float and potentiometer arrangement could be used to provide depth information in analogue form.

Sensors which are inherently analogue – perhaps providing information in the form of variable resistance – may still be used to provide digital data by incorporating the transducer into a *comparator* circuit. In this case, one input of the comparator is connected to a constant voltage reference. The other input is then connected to a voltage divider, formed by the sensor and a fixed resistor.

Quantity	Sensor/actuator
Light intensity	Input – light dependent resistor (LDR), photo diode/transistor, solar cell
	Output – lamp, shutter.
Temperature	Input – thermistor, thermocouple, pyrometer, thermostat, pyroelectric sensor
	Output – heater (resistor), fan
Pressure	Input – diaphragm
	Output – pump, valve
Stress/strain	Input – strain gauge, piezo-electric crystal
Force/weight	Input – load cell, piezo-electric crystal, spring
Sound	Input – microphone, piezo-electric crystal
	Output – loudspeaker, piezo-electric crystal
Rate of flow	Input – venturi, pitot tube, turbine, doppler effect sensor
	Output – pump, valve
Attitude/vibration	Input – mercury tilt switch, vibration switch (mercury), gyrosope
Position	Input – potentiometer, Gray-coded disc, linear variable differential transformer (LVDT), proximity sensor (capacitive/inductive/magnetic), slotted/reflective opto-coupler, infrared beam, microswitch, reed switch
	Output – motor, solenoid
Depth	Input – float activated switch, potentiometer (and float), proximity sensor
	Output – valve, pump
Speed	Input – tacho-generator, slotted/reflective opto-coupler, doppler effect sensor
	Output – motor, brake
Magnetic field	Input – Hall effect sensor
	Output – electromagnet

Table 7.1 Common sensors and actuators.

By suitable calibration, the comparator may be used to provide a digital indication that some analogue quantity has passed through a threshold – possibly signalling an alarm condition. A typical example would be the use of an LDR to provide a digital indication of light level, as shown in figure 7.3.

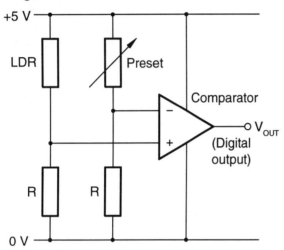

Fig. 7.3   Digital light threshold sensor.

The same type of sensor may be placed in an *instrumentation amplifier* circuit to provide an analogue indication of the measured quantity, as shown in figure 7.4.

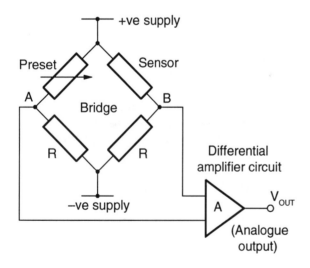

Fig. 7.4   Instrumentation amplifier circuit.

By adjusting the preset so that it has the same resistance as the sensor, the voltage at points A and B may be made equal, causing the bridge to be 'balanced'. Any deviation in the resistance of the sensor will then cause the bridge to become unbalanced, resulting in a potential difference at the amplifier input terminals. This voltage difference may then be amplified by a differential amplifier, before being converted into a form which can be interpreted by a computer.

As mentioned previously, computers operate internally on digital information and are not directly compatible with analogue data. Special *digital to analogue* and *analogue to digital* converter ICs are therefore used to convert information to or from analogue form, as shown in figure 7.5.

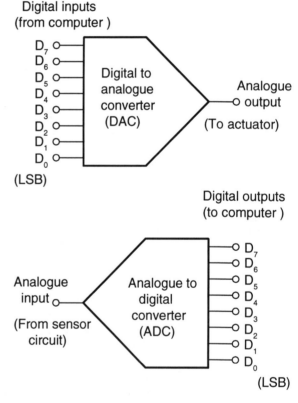

Fig. 7.5   Converting between analogue and digital data types.

Digital to Analogue Conversion

Most commonly available digital to analogue converter ICs contain an *R-2R ladder network*, such as that shown in figure 7.6.

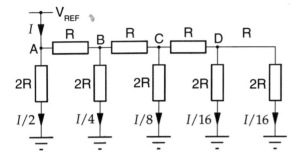

Fig. 7.6 R-2R ladder network.

Although it is not obvious at first sight, this network has the property that it produces binary weighted currents in each of the vertical branches.

This is based on the fact that at any node (A, B, C or D), the resistance seen looking 'to the right' is equal to '2R'. As this resistance appears in parallel with the vertically oriented resistor connected to each node, the current flowing into each node (from the left) divides equally between the two branches. Thus a current of I amperes flowing into node A splits into two currents of $I/2$ amperes, and this process of current division occurs at each node in the network, moving to the right.

In a practical digital to analogue converter, these binary weighted currents are selectively passed to an operational amplifier circuit using FET-controlled *analogue switches*, as shown in figure 7.7 overleaf.

Practical Exercise   R-2R Network

Circuit Diagram

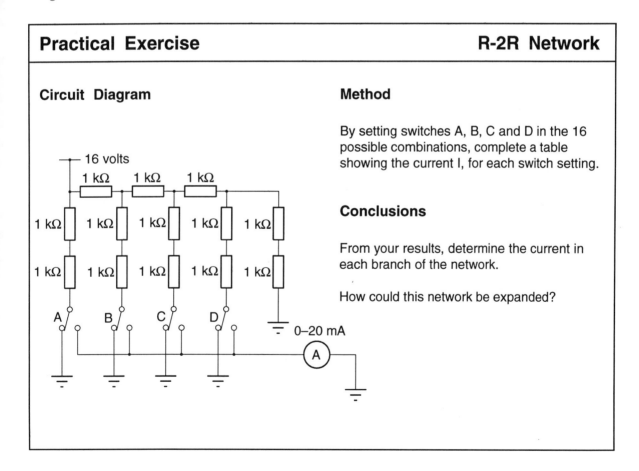

Method

By setting switches A, B, C and D in the 16 possible combinations, complete a table showing the current I, for each switch setting.

Conclusions

From your results, determine the current in each branch of the network.

How could this network be expanded?

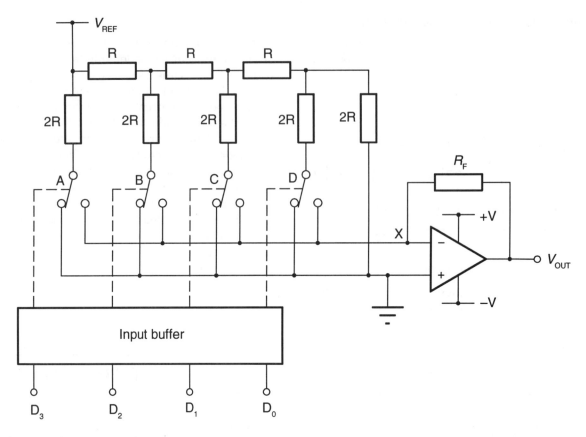

Fig. 7.7   Digital to analogue converter.

Although a detailed analysis of the circuit operation will not be performed here, the following points should be noted.

- The voltage at point 'X' is approximately 0 volts, due to the *virtual earth* principle.
- Due to the high input impedance of the operational amplifier, almost all current flowing into node X from the left also flows through the feedback resistor R_F. The output voltage therefore depends on the current flowing into node X (and hence on the state of the analogue switches), and on the value of R_F.
- Varying the reference voltage V_{REF} alters the full scale output voltage.
- Due to the use of an *inverting amplifier* circuit, the output voltage and the reference voltage are of opposite polarity.

There are two basic methods of interfacing a digital to analogue converter to a computer. For experimental purposes, the simplest approach is to connect the digital inputs directly to an output port. With an 8 bit converter, an entire port would be required, as shown in figure 7.8.

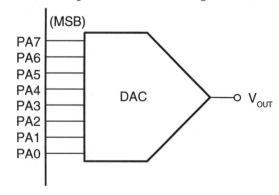

Fig. 7.8   Output port interfacing approach.

This is a rather wasteful approach if the converter IC is to permanently connected to the computer, as the output port is not available for other purposes.

An improved technique is to design suitable interfacing circuitry so the converter appears as either a memory location (*memory mapped input/output*) or as an output port address (*IO mapped input/output*).

Digital to analogue converters which are intended for direct connection to the microprocessor data bus are normally fitted with an input latch and a latch enable input. The interfacing circuitry must then cause the latch enable pin to be pulsed as valid data is present on the data bus. As an example, figure 7.9 shows pin connections of the commercially available ZN428E 8 bit D/A converter IC, which is intended for use with microprocessor-based systems.

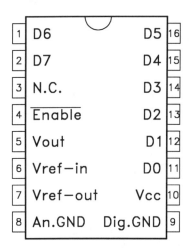

Fig. 7.9 ZN428E 8 bit D/A converter IC.

Notice here that an internal voltage reference (2.55 volts) is provided on pin 7. In normal use, this is directly connected to pin 6. It is also possible to connect externally generated reference voltages, thus altering the maximum output voltage of the converter.

Assuming the maximum output voltage is set to 2.55 volts, the 256 possible input codes produce output voltages which increment in steps of 0.01 volts (0.00–2.55 volts).

Analogue to Digital Conversion

Analogue to digital conversion may be performed using a range of techniques, including the

- counter-ramp technique,
- tracking converter,
- dual slope converter,
- successive approximation converter,
- flash converter.

An understanding of these conversion methods is not required if a commercially available converter IC is to be used. However some appreciation of the advantages and disadvantages of each approach may prove useful when selecting a device for a particular application. (Readers who require a detailed explanation of the various conversion methods may consult any of the large range of electronics text books which are available.)

It will be shown here that a simple analogue to digital converter may be constructed from an existing digital to analogue converter, a comparator and a suitable software algorithm. This approach allows a single converter IC to be used to input or output analogue information, although only one direction of data transfer is possible at any time. This may prove useful in cost-sensitive applications.

The basic circuit arrangement is shown in figure 7.10.

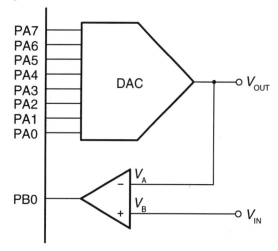

Fig. 7.10 Simple analogue to digital converter.

Practical Exercise

Waveform Generation

Method

Write software to generate ramp, square and sine waveforms using a digital to analogue converter IC.

Conclusions

Use an oscilloscope to observe the output waveforms produced by the program.

By increasing the vertical scale, attempt to identify the 'stepped' nature of the output waveform. How could this effect be removed?

Example Algorithm

a) Ramp

    1 Initialise port as an output
    2 count = 0
    3 Send count to port
    4 Time delay
    5 count = count + 1
    6 Goto step 3 always

Hint (Sine wave)

Use a *look-up table* to store successive sine wave values in memory. These should be calculated assuming the sine wave has a midpoint value of 128_{10} and an amplitude of 127_{10}. Calculate values for the table in 15° increments.

Use *indexed addressing* to read each table entry.

Required Output Waveforms

a) Ramp

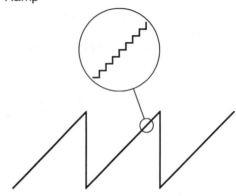

b) Square

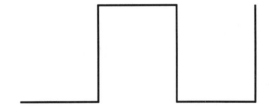

c) Sine

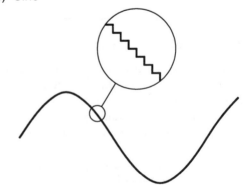

When used as an analogue to digital converter, the unknown analogue input is applied to V_{IN}, while V_{OUT} is unused.

The comparator is connected to both the output of the digital to analogue converter (V_A), and to the unknown input voltage (V_B) and provides an indication of which of these two signals is the greater. (If $V_A > V_B$ then the comparator output is logic 0, while the output is logic 1 if $V_B > V_A$.)

Using the very basic information provided by the comparator, it is possible to develop a range of algorithms to determine the unknown input voltage. The simplest algorithm is the *counter-ramp* technique which is shown below.

1   Initialise port A as an output, and port B as an input.
2   count = 0.
3   Send count to port A.
4   Test comparator output (read port B and mask unwanted bits).
5   If comparator output = 0 then goto step 8 (conversion complete).
6   count = count + 1.
7   Goto step 3 always.
8   Return last value output (as the result) and end.

Figure 7.11 gives typical waveforms produced during the conversion process.

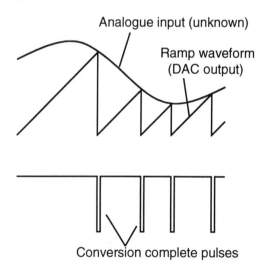

Fig. 7.11   Counter-ramp conversion method.

(The conversion waveforms shown in figure 7.11 assume the process of analogue to digital conversion is running continuously.)

An obvious disadvantage here is that the time taken to perform the conversion depends on the magnitude of the unknown analogue voltage, with larger values resulting in longer conversion times. An improved method is the *tracking converter*, whose operation is shown in figure 7.12.

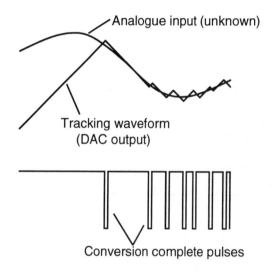

Fig. 7.12   Tracking converter waveforms.

With this method, the comparator is first tested to determine whether the unknown analogue input is greater than or less than the converter output voltage. The count value sent to the digital to analogue converter is then incremented or decremented, as appropriate, until the comparator output changes sign.

The tracking method is more efficient than the simpler counter-ramp approach, assuming the converter operates continuously. This can be seen by comparing the number of conversion complete pulses in figure 7.12, compared with figure 7.11. (It will be left as an exercise for the reader to develop an algorithm for this conversion method.)

Another approach frequently used in commercially available converter ICs is the *successive approximation* approach, which is illustrated in figure 7.13 overleaf.

Practical Exercise   Analogue to Digital Conversion

Method

Write programs to perform analogue to digital conversion based on

- the counter-ramp technique,
- the tracking converter,
- the successive approximation method.

Results and Conclusions

Use an oscilloscope to identify typical conversion waveforms for each technique.

Compare the effectiveness of each method.

Circuit Diagram

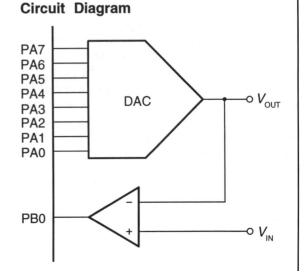

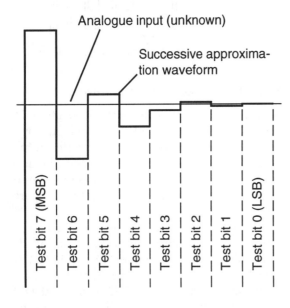

Fig. 7.13 Successive approximation method.

As the name suggests, the successive approximation method uses a series of 'guesses' to identify the unknown analogue voltage with increasing precision. One step is required for each bit of the converter (so an 8 bit converter requires eight steps).

The process begins by setting the most significant bit of the converter and then testing the comparator output to determine whether the unknown voltage is greater than or less than the first approximation.

Depending on the result, the tested bit is either left or removed. If the unknown voltage is higher then the bit remains set, while if it is lower the tested bit is cleared.

This process is then repeated for each of the converter bits in turn, until the output voltage approximates as closely as possible to the unknown input signal.

Successive approximation is one of the more efficient conversion techniques and is commonly used by commercially available devices. (Very high speed conversion is normally performed by *flash* converter ICs, which are not considered here.)

In commercially available converter ICs, the conversion process is normally started by pulsing the *start conversion* pin, while the *conversion complete* pin indicates that the digital outputs are valid. (Microprocessor compatible devices also possess a latch with tri-state outputs.)

Practical Exercise   Signal Processing

Method

Develop software to record and process data that has been applied to the input of an analogue to digital converter.

a) Data logging

Write software to read 100 samples from the converter and store them in successive memory locations.

b) Computer oscilloscope

If your computer has a graphics facility, write software to display the recorded data as a waveform on the computer screen.

c) Digital filtering

Write software to filter the data, using the following (averaging filter) algorithm.

$$F_N = \frac{D_N + D_{N-1} + D_{N-2} + D_{N-3}}{4}$$

Where

F_N = filtered data (Nth sample)
D_N = raw data (Nth sample)

d) Derivative and integral

By subtracting successive samples, determine the *rate of change* of the data. By adding samples to a running total, calculate the *integral* of the sampled data.

Figure 7.14 shows the pin connections of the commercially available ADC0804 8 bit A/D converter IC.

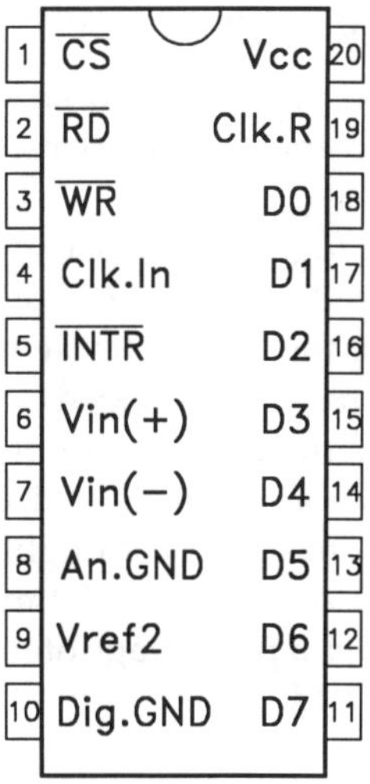

1	$\overline{CS}$	Vcc	20
2	$\overline{RD}$	Clk.R	19
3	$\overline{WR}$	D0	18
4	Clk.In	D1	17
5	$\overline{INTR}$	D2	16
6	Vin(+)	D3	15
7	Vin(−)	D4	14
8	An.GND	D5	13
9	Vref2	D6	12
10	Dig.GND	D7	11

Fig. 7.14 ADC0804 A/D converter IC.

The addition of a small number of passive components is all that is required to allow this device to operate with most microprocessor types. Notice that read, write, chip select and interrupt request (conversion complete) pins are all provided.

The input may also be scaled by applying a suitable voltage to pin 9 (normally derived from V_{CC} using a voltage divider arrangement). The maximum allowed input voltage is given by the voltage on pin 9 multiplied by two.

As was the case with digital to analogue converters considered previously, these devices may be interfaced using one or more input ports, or as a memory mapped (or IO mapped) peripheral. The first method should only be used for experimental purposes due to the large number of port bits which are tied up (10 at least).

An interrupt-driven approach is also preferred with slower types of analogue to digital converter, because polling routines are wasteful of the microprocessor's time.

When converting data to or from analogue form, several points should be noted.

- Unlike a truly analogue signal, the output of a digital to analogue converter is not continuously variable. Instead, the output voltage is allowed to take a discrete number of levels. (A converter with N bits is capable of producing 2^N output voltage levels).
- When an analogue input is converted to digital form, the signal is transformed into the nearest appropriate digital code. This process inevitably leads to small 'rounding-off' or *quantisation* errors.
- Analogue signals are only sampled (or produced) at discrete intervals of time.
- The sampling frequency when inputting analogue data should be at least twice as high as the highest frequency component in the input data, if *aliasing errors* are to be avoided. The input data may be *band-limited* by passing it through a suitable *low-pass filter*, prior to conversion to digital form.
- If an analogue signal changes during the conversion process, this may lead to errors, since the converter will attempt to 'chase' the input signal. This may be avoided by the use of a *sample and hold* circuit, which uses a capacitor and analogue switch arrangement to 'freeze' the analogue input, prior to conversion.

A typical sample and hold circuit is shown in figure 7.15.

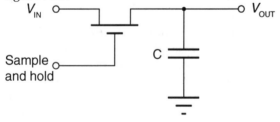

Fig. 7.15   Sample and hold circuit.

D.C. and Stepper Motors

In the first volume, basic circuits were introduced for controlling external devices such as lamps, d.c. motors and solenoids. At that time discussion was limited to simple on/off control methods. The following section shows how *open loop* and *closed loop* techniques may be used to control the direction, speed and position of d.c. and stepper motors.

Speed Control of D.C. Motors

Small d.c. (6/12 volt) motors may be directly switched from an output port using a circuit such as that given in figure 7.16.

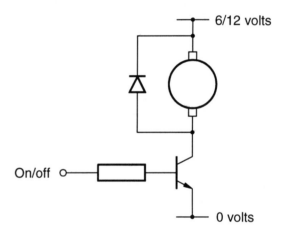

Fig. 7.16   Simple d.c. motor control circuit.

(A Darlington transistor may be used where higher current gain or increased current rating is required.)

A continuous logic 1 or logic 0 applied to the input may be used to turn the motor on or off respectively, but the same circuit may also be used to control the motor's speed.

By repeatedly turning the motor on and off, the speed of the motor may be varied between 0 and 100% of the maximum. Roughly speaking, the proportion of 'on' time to 'off' time is equal to the motor speed, as a percentage of maximum, as shown in figure 7.17.

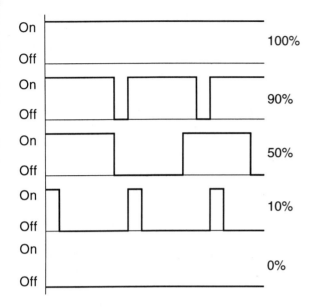

Fig. 7.17 Pulse width speed control.

1 Initialise port as an output.
2 count = 0, motor = 1, threshold = ?.
3 Send motor state to port.
4 count = count + 1.
5 If count = threshold then motor = 0.
6 If count = 0 then goto step 2.
7 Goto step 3 always.

The simplest way to implement *pulse width* speed control is to use a counting routine, in which the motor is turned on as the count passes zero and then off at some threshold. It the selection of this threshold value which determines the proportion of on to off time, as shown by the following algorithm.

Bi-directional d.c. motor control is also possible, but the direction of current flow through the motor must be reversed in order to change the motor's direction of rotation. The bridge circuit of figure 7.18 may be used in this case.

Notice here that complementary transistors (NPN and PNP) are used in each branch, thus for a given applied logic level, one of the transistors will always be enabled, while the other is disabled. By applying opposite logic levels to inputs A and B, the motor direction may be controlled, as shown below.

A	B	Motor function
0	0	Off
0	1	Forward
1	0	Reverse
1	1	Off

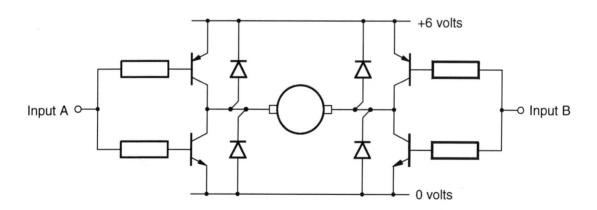

Fig. 7.18 Bi-directional d.c. motor control circuit.

Similar techniques may be used to interface the coils of a stepper motor to an output port, as shown in figure 7.19.

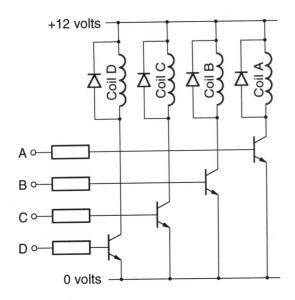

Fig. 7.19 Stepper motor interface circuit.

In this case, each of the stepper motor coils is controlled by a different output port bit. The stepper motor may then be made to step forwards or backwards by outputting an appropriate control sequence, as shown below.

A	B	C	D	Result
0	1	1	0	Forward
0	0	1	1	Forward
1	0	0	1	Forward
1	1	0	0	Forward
0	1	1	0	Forward
etc.				

A	B	C	D	Result
0	1	1	0	Backward
1	1	0	0	Backward
1	0	0	1	Backward
0	0	1	1	Backward
0	1	1	0	Backward
etc.				

These sequences consist of the binary pattern 0110_2 which is rotated to the right to cause forward movement, or to the left to move the stepper motor in reverse.

Each step causes the stepper motor to rotate through a certain angle (7.5° for example). By maintaining a running total of the number of pulses output, the total angle through which the stepper motor has turned may be determined.

It is also possible to vary the speed of the stepper motor by altering the time delay between output pulses. This ability to accelerate or decelerate a stepper motor can be particularly useful when driving loads with high rotational inertia.

Finally, it should be noted that a range of ICs is commercially available, which greatly simplify the interfacing of motors to TTL/CMOS compatible logic. *Fully controlled bridge* ICs are available which allow d.c. and stepper motors with current ratings of several amperes to be directly controlled from a microcomputer output port.

The SAA1027 stepper motor controller IC is also popular. This IC contains a counter and code converter which automatically drives the outputs in the correct sequence, while only two input signals are required. The direction of rotation is selected by applying the appropriate logic level to pin 3, while a low to high transition on pin 15 causes a single step to occur in the selected direction. (These inputs are not directly TTL compatible, so an interface circuit is required.)

Closed Loop Control

The motor control methods considered above are referred to as *open loop* because no feedback is taken from the output.

In *closed loop control*, some kind of sensor is connected to the output, which allows the actual output (speed, position, temperature etc.) to be compared with the desired output. If an error is detected, then corrective action can be taken by the control program. Closed loop control is inherently more accurate than open loop control for this reason. Figure 7.20 shows a block diagram of a typical closed loop system.

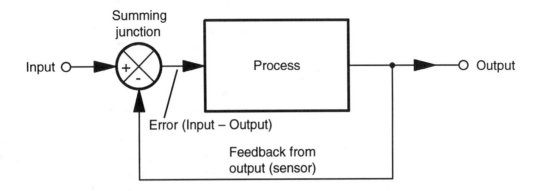

Fig. 7.20 Typical closed loop block diagram.

The type of feedback provided will depend on the nature of the control system. With a *position control system*, a variable voltage might be produced by a potentiometer, depending on the position of the output shaft. This could then be input to the computer using an analogue to digital converter, as shown in figure 7.21.

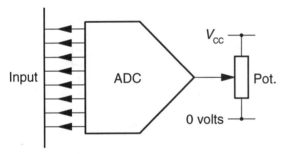

Fig. 7.21 Position control system feedback.

Similarly, in a *speed control system*, a *tacho-generator* could be driven from the motor shaft. The variable voltage produced could then be fed back to the computer, after conversion into digital form.

It is also possible to provide feedback of actual position or speed, directly in digital form. For example, a *Gray-coded disc* could be used in conjunction with *reflective* or *slotted opto-couplers*, to provide a digital indication of position, as shown in figure 7.22.

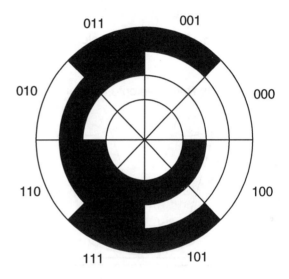

Fig. 7.22 3 bit Gray-coded disc.

(Gray code has the property that only one bit changes at any time, thus eliminating the possibility of unwanted transient codes caused by several bits changing at slightly different times, as the disc rotates.)

A *slotted disc* could also be used to produce a single pulse following each revolution of a drive shaft, which would be useful in a speed control system. In this case, a slotted opto-coupler would normally be used. Figure 7.23 (overleaf) shows the two available types of opto-coupler.

Slotted disc

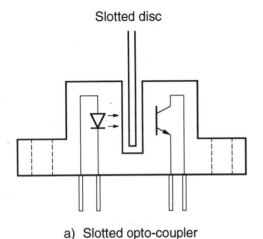

a) Slotted opto-coupler

b) Reflective opto-coupler

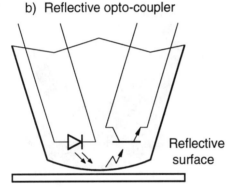

Reflective
surface

Fig. 7.23   Reflective and slotted opto-couplers.

Opto-couplers consist internally of an LED and a *photo-transistor*, or light-activated transistor. Energising the LED causes the transistor to be switched on, which then provides a current or voltage signal in the receiver circuit. (Recall the 'transistor as a switch', which was covered in *Microelectronic Systems – Book One.*)

The lack of any electrical connection between the input and output, provides electrical isolation of several thousand volts between the input and output section of the device. This important safety feature is often built into interface circuitry in the form of *opto-isolator* ICs, whose only function is to break the electrical connection between the low power input and high power output circuits.

Using Opto-Isolators

Optical isolator ICs may be used to switch a range of electronic devices including,

- transistors,
- triacs,
- triacs with in-built zero-crossing detection.

These devices provide the electronic designer with a simple and relatively safe method of controlling d.c. and a.c. power equipment. They are also useful when data must be input in electrically noisy environments. In each case, it is the *electrical isolation* which is the important factor!

Figure 7.24 illustrates the use of an opto-transistor isolator to input information from a switch.

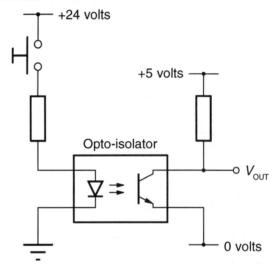

Fig. 7.24   Reading a switch in an electrically-noisy environment.

Notice here the use of separate electrical grounds in the input and output sections of the circuit, which acts as a barrier against (conducted) electrical noise.

Opto-triac isolators are also available which allow a.c. powered equipment to be controlled. These optically-coupled triacs typically have a 400 volt rating (suitable for direct mains connection) and a maximum current of 100 mA. Low power devices may be directly connected, as shown in figure 7.25.

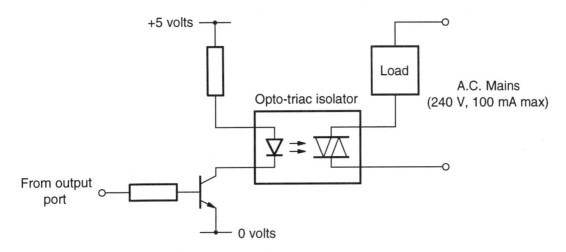

Fig. 7.25 Typical opto-triac isolator circuit.

With higher powered loads, the opto-triac may be used to provide the gate pulse to another triac, via a current limiting resistor, as shown in figure 7.26.

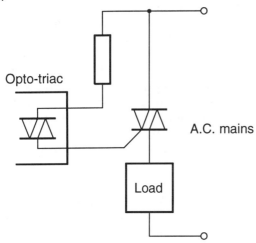

Fig. 7.26 Controlling high power loads.

A brief pulse applied to the input terminal causes the opto-triac to be triggered. The triac then conducts until the end of the current cycle of the mains, at which point it automatically turns-off, or *commutates*. (Unlike an optically coupled *thyristor*, a triac is capable of conducting in both halves of the mains cycle.)

There are two basic strategies which may be used when controlling the power supplied to an a.c. load. These are *phase control* and *integral cycle* (or *burst fire*) control.

In phase control, the trigger pulse is applied at some fixed time following the zero cross-over, as shown in figure 7.27.

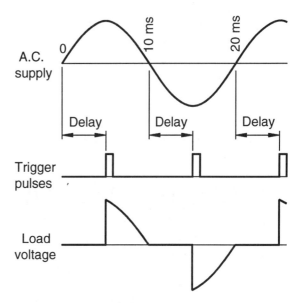

Fig. 7.27 Typical phase control waveforms.

By varying the time delay between 0–10 ms, the power supplied to the load may be controlled between zero and 100%.

Clearly, the generation of time delays in the range 0–10 ms is easily achieved using a microcomputer, but in this case, some way of identifying the *zero volt cross-over* point must be found. One solution is to drive the input of a comparator from a low voltage secondary transformer winding, as shown in figure 7.28.

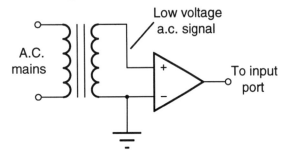

Fig. 7.28 Simplified zero cross-over detector.

Assuming that zero cross-over information is available from an input port bit, and the a.c. load is controlled from an output port bit (via an opto-triac isolator), a possible phase control algorithm is given below.

1 Initialise ports.
2 Wait for zero cross-over (0–1, or 1–0).
3 Time delay (0–10 ms).
4 Fire triac (brief pulse).
5 Goto step 2 always.

The second method, known as *integral cycle* or *burst fire* control, involves turning the load on and off for several complete cycles at a time, as shown in figure 7.29.

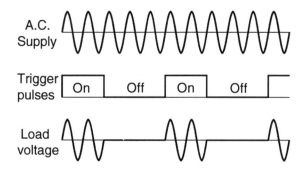

Fig. 7.29 Integral cycle control waveforms.

The power supplied using this method is controlled by the ratio of cycles where the load is 'on' or 'off'. When using a computer as the controller, two different time delays could be used, as shown in the following algorithm.

1 Initialise port as an output.
2 Enable triac (output 1 to port).
3 Time delay (number of 'on' cycles multiplied by 20 ms).
4 Disable triac (output 0 to port).
5 Time delay (number of 'off' cycles multiplied by 20 ms).
6 Goto step 2 always.

One problem with this method is that the control pulses are not actually synchronised with the zero cross-over points of the mains supply! This means that the load may be energised part of the way through a mains cycle. Similarly, when the enable pulse is removed, the triac will continue to conduct until the end of the current cycle.

The sudden change in load current caused by this lack of synchronisation tends to generate electrical noise (which is also a disadvantage of the phase control method considered earlier). For this reason, optically coupled triacs are available with built-in *zero-crossing detectors*. When these opto-couplers are enabled, the triac is actually triggered at the start of the following mains cycle (when the current is zero), which minimises the generation of electrical noise.

In general, integral cycle control is more suitable for loads with slower response times, such as large heaters. Phase control is ideal in time-critical applications, such as light dimming, where significant flicker would be a disadvantage.

Practical Exercise   A.C. Power Control

Introduction

Based on the material in this section, use an optically-isolated triac to control the power supplied to a low power a.c. load, such as a 12 volt (1.2 watt) lamp.

Develop phase control and integral cycle (burst fire) control software, together with suitable hardware for each technique.

(A separate zero-crossing detector will be required for the phase control method, while this feature may be built-into the opto-isolated triac used for burst fire control.)

Results and Conclusions

Record typical load voltage waveforms for both control strategies (at low, medium and high power).

Compare the effectiveness of the two methods for controlling the power supplied to the load. Hence suggest suitable applications in each case.

Basic circuit

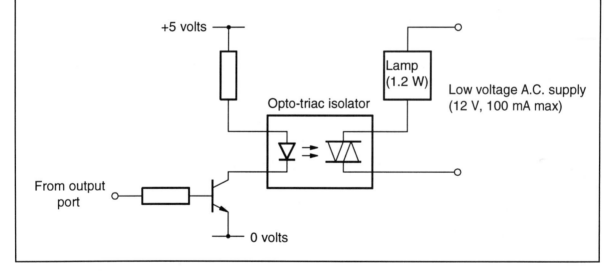

Summary

The main points covered in this chapter are:

- The use of *multiplexing* with 7-segment displays results in lower component costs and a reduced number of computer connections, although the software complexity is increased. Each digit is illuminated singly in a repeating sequence, giving the appearance of continuous illumination.
- *Transducers* or *sensors* allow analogue quantities such as light intensity, temperature or pressure to be measured. These are often used in conjunction with a suitable *instrumentation amplifier* circuit, which produces a variable analogue voltage as the measured property alters.
- Computers operate internally on digital signals, so an analogue value must be converted into digital form before being input. This is performed by an *analogue to digital converter* IC.
- A computer may also produce analogue voltages by using a *digital to analogue converter* IC.
- The *R-2R* network is the most commonly used method of generating an analogue signal from a series of binary weighted digital inputs.
- Analogue to digital conversion can be performed in software using a digital to analogue converter and a comparator. Several software-based algorithms may be used including the *counter-ramp*, *tracking* and *successive approximation* methods. Some of these approaches are also used in commercially available converter ICs (together with the *dual-ramp* and *flash* conversion methods).
- The ability to generate analogue signals may be used in waveform generation, and also allows analogue properties such as speed, temperature or position to be controlled.
- The speed of a d.c. motor may be controlled using a digital (on/off) signal by producing a variable pulse width control waveform.

- Directional control of d.c. motors is possible using a *bridge* circuit. In this case, two control inputs are used. (It is also possible to combine directional and pulse width speed control techniques, with the same circuit.)
- Stepper motors may be controlled by directly switching the motor coils using transistor switching techniques. Commercially produced motor control ICs are also available which may be used with d.c. and stepper motors.
- In control systems, greater accuracy is achieved by using *closed loop* (rather than *open loop*) techniques. This is based on the use of feedback, which allows the actual output to be compared with the desired output. In the event of an error, the control program is then able to take corrective action.
- Feedback from a control system may be either analogue or digital. Examples of analogue feedback include potential dividers and tacho-generators (which may be used to indicate position or speed respectively). The same information could be provided in digital form using opto-couplers in conjunction with Gray-coded and slotted discs respectively.
- Opto-isolators allow devices such as transistors and triacs to be controlled from an output port. Their main advantage is the high *electrical isolation* between input and output, which is an important safety feature.
- Opto-triac isolators may be used to control the power supplied to an a.c. load using either *phase control* or *integral cycle* (*burst fire*) techniques.
- In phase control, the computer must be aware of the *zero cross-over point*, as this is the reference from which the firing time delay is calculated.
- Opto-coupled triacs with *zero-crossing detectors* are normally used in integral cycle control, which greatly reduces the generation of electrical noise.

8   Integrative Assignments

Aims

When you have completed this chapter you should be able to:

1   Successfully complete an assignment that contains objectives taken from at least three related BTEC units.
2   Develop and enhance a range of transferable skills that are essential in working life.
    These Common Skills cover abilities such as personal development, communicating, numeracy, working with others, use of technology, problem solving and design skills.

Integrative Assignments

A *programme of integrative assignments* (PIA) is an essential part of BTEC programmes of study at National level. This chapter introduces two integrative assignments which could form part of an engineering-related course with an electrical bias.

By definition, an integrative assignment must contain identified objectives from at least three of the units being studied on the programme. These assignments are intended to enable students to recognise the transferrable nature of technical knowledge, and to encourage the development of personal skills which may be used throughout working life.

The assessment of BTEC National level courses is based on the attainment of objective-related competences, and also on seven different personal skills. Example common skills claim forms are included with each assignment in this chapter.

Control Systems

The purpose of this assignment is to develop and evaluate different ways of controlling the behaviour of typical engineering systems.

It is intended that the reader should build and test a range of open loop and closed loop systems, based on the information given here. The advantages and disadvantages of each approach may then form the basis of a report, while the solutions developed may be used in future design work.

(The treatment of control systems given here is deliberately 'non-mathematical'.)

Open Loop Control

The actual output produced by an open loop system is not tested, which can lead to poor accuracy. An example of an open loop system would be a cooling fan which was permanently switched on, regardless of the temperature of the system (such as a computer cooling fan).

This type of system suffers from several disadvantages, due to the lack of temperature feedback.

● The fan runs constantly which wastes energy.
● Ambient temperature is not taken into account, nor is the power dissipated by the system (which may vary).
● Although the temperature will not exceed the maximum allowed, it may vary over a very wide range.

A simple open loop cooling system is shown in figure 8.1.

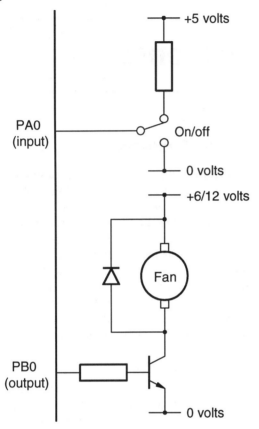

Fig. 8.1 Open loop cooling system.

A simple control program could be developed for this system, as shown by the following algorithm.

1 Initialise ports (A = input, B = output).
2 Read port A (into accumulator).
3 Output accumulator to port B.
4 Goto step 2 always.

An improvement on this system would be to use a series of switches connected to port A to indicate the required motor speed. Pulse width speed control techniques could then be used to vary the current supplied to the motor, and hence its speed. A possible circuit diagram is shown in figure 8.2.

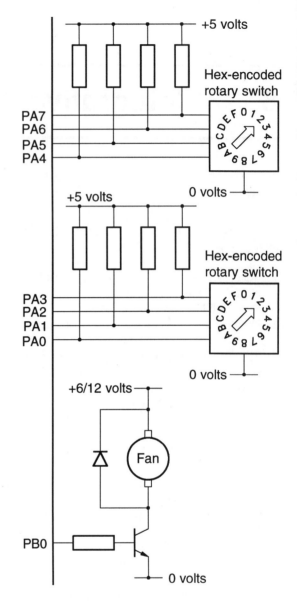

Fig. 8.2 Proportionally controlled open loop cooling system.

The switching arrangement may be simplified by the use of hex-encoded rotary switches, which also enables the input code to be read easily.

A possible pulse width speed control algorithm is given overleaf, which uses the value obtained from the switches as the switching threshold.

1. Initialise ports (A = input, B = output).
2. Read threshold from port A.
3. Count = 0, motor = 1.
4. Send motor state to port B.
5. Count = count + 1.
6. If count = threshold then motor = 0.
7. If count = 0 then goto step 2.
8. Goto step 4 always.

Although the required speed of the motor may now be set by the operator, it will be found that the actual speed is not controlled very precisely, and that the speed may decrease if the load driven by the motor increases. (Try this.)

Simple Closed Loop Control Methods

In closed loop control, the actual output produced (speed, position, temperature etc.) is compared with the required output. Any error causes the controller to take compensating action, thus returning the system to the desired state. Figure 8.3 shows the addition of a temperature sensor and analogue to digital converter to the previously considered temperature control system.

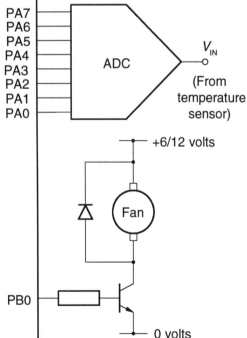

Fig. 8.3 Closed loop temperature control system.

Various control strategies are now possible. The simplest (and crudest) technique is to switch on the fan whenever the temperature rises above a pre-determined threshold. A possible algorithm is given below.

1. Initialise ports (A = input, B = output)
2. Threshold = ? (Set in software).
3. Read temperature from sensor.
4. If temperature > threshold then fan = 1 (on), else fan = 0 (off).
5. Send fan state to port B.
6. Goto step 3 always (step 2 if threshold can be altered while the program runs).

This single threshold technique works quite well but the fan tends to switch on and off every few seconds as the temperature hovers around the threshold. This may be a distraction (particularly in the case of a computer cooling fan), and may tend to increase component wear. An improved technique is use two thresholds, as shown in figure 8.4.

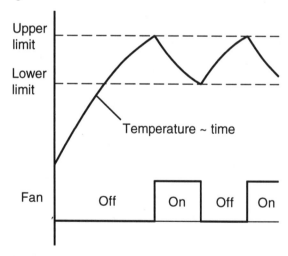

Fig. 8.4 Dual threshold closed loop control.

As the temperature rises above the upper temperature threshold, the fan is turned on. It continues to run until the temperature has fallen below the lower threshold, at which point the fan is switched-off. This is a considerable improvement over the single threshold technique, as the fan switches on and off less frequently.

(The development of an algorithm for this control strategy is left as an exercise for the reader.)

One disadvantage of this technique is that the output variable (temperature in this case) is only loosely controlled. With a closed loop *speed control system* or *position control system* for example, it may be necessary to control the output variable more precisely.

Another problem is that the fan is either switched on or off, which causes rapid changes in temperature. Varying the speed of the fan (using pulse width speed control) would allow abrupt changes in temperature to be avoided.

PID Control Methods

A range of feedback types is used in more demanding control applications including *proportional, integral* and *derivative* (*PID*) terms. PID controllers are commercially available (based either on operational amplifier circuits or computer controllers) which allow these three type of feedback to be mixed in various proportions in order to obtain the optimum degree of control.

In *proportional control*, the error between the actual and required output is measured. A compensation is then applied to the system, whose size and polarity depends on the magnitude and sign of the error. For example, if the motor speed in a position control system is found to be too low, the input to the motor drive circuitry will be increased by an appropriate amount. As the motor speed approaches that desired, the error size diminishes towards zero.

An obvious question is how much compensation should be applied to the speed control circuitry via the feedback network?

The general expression for a closed loop system which uses negative feedback is given in equation 8.1

$$\text{Gain} = \frac{\text{Output}}{\text{Input}} = \frac{A}{1 + A\beta} \qquad (8.1)$$

where A is the gain of the controller section, and β is the gain of the feedback network, as shown in figure 8.5.

Assuming that $A = 1$ and that β is much less than 1, the overall gain of the closed loop system will approach unity. (Of course if $\beta = 0$ then the system becomes open loop once again!)

For example, assume that $\beta = 1/16$, and the input is set to 128_{10}. From equation 8.1, the gain of the system will be 0.941, so for an input of 128_{10}, the output should settle at around 120_{10}. (An output of 120_{10} will produce a feedback signal of 8_{10}, which is then subtracted from the original input, giving a drive signal to the motor speed controller of 120_{10}.)

When a computer is used as the controller, feedback may be supplied by a tacho-generator, which will produce a voltage proportional to the speed of the motor. This may then be converted to digital form before being input by the computer. The speed of the motor may be controlled using pulse width speed control techniques, with the width of the drive pulses being dynamically altered according to the size of the error signal.

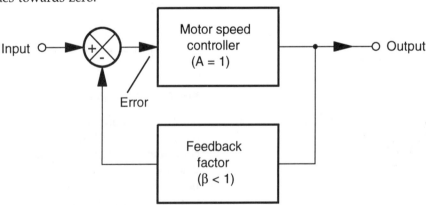

Fig. 8.5 Motor speed control system which uses negative (proportional) feedback.

A possible algorithm for a motor speed controller which uses proportional feedback is shown below.

1. Initialise ports.
2. Feedback factor β = ? (set in software).
3. Input = ? (set in software).
4. Read actual motor speed from sensor.
5. Multiply actual speed by feedback factor (β << 1).
6. Threshold = Input − feedback
7. Output one (or more) pulse cycles, using the above switching threshold.
8. Goto step 4 always (step 2 if β can be varied while the program runs).

A simple way to implement step 5 is to shift the value obtained from the analogue to digital converter to the right, which effectively halves the number. In the example given on the previous page, division by 16 could be achieved, either by four shift instructions, or by swapping nibbles in the accumulator.

One consequence of using small values for β, is the loss of accuracy that occurs when the least significant bits of the feedback signal are discarded during division. (When β = 1/16, the lower nibble is discarded.) A small error in the output speed might not cause a change in the feedback signal, which means that the controller will only take corrective action when the error signal is of a significant size.

The fact that a small constant error can exist at the output is a disadvantage where proportional feedback is used by itself. For this reason, *proportional* and *integral* feedback techniques are often combined to provide greater output accuracy.

By definition, the integral of a signal is the 'area under the curve'. With a constant input of X volts, the integral is given by

$$\int_0^t X \,.\, dt = Xt + \text{constant} \qquad (8.2)$$

If the voltage applied to an integrator is actually a constant error signal produced by a proportional controller, it can be seen that the integral will be a ramp voltage, as shown in figure 8.6.

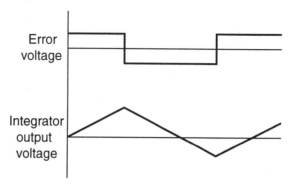

Fig. 8.6 Integration of a d.c. error signal.

The effect of combining proportional and integral feedback types is to prevent a d.c. error from persisting for any length of time.

With pulse width motor speed control, the integral of the output error could be used to modify the switching threshold. For example, if the actual motor speed was too high, the error signal would be negative. The integral controller would in this case decrement the threshold value until the error signal has disappeared.

An algorithm for a simple integral controller is given below.

1. Initialise ports (A = input, B = output).
2. Desired speed = ? (set in software).
3. Initial threshold = desired speed.
4. Read actual motor speed from sensor.
5. If actual speed < desired speed then threshold = threshold + 1.
6. If actual speed > desired speed then threshold = threshold − 1.
7. Output one (or more) pulse cycles, using the above switching threshold.
8. Goto step 4 always .

(It is left as an exercise for the reader to produce an algorithm that combines proportional and integral techniques.)

A third type of signal which is often used in control systems is *derivative* feedback. In this case, the size and polarity of the signal depends on the *rate of change* of the output.

To understand why derivative feedback may sometimes be required, consider as an example the motor speed controller, where the required motor speed is 10% of the maximum.

The friction of a stationary motor is normally greater than when the motor is running, so it may be found that a proportional controller is unable to start the motor at such low speeds. The addition of integral control will enable the motor to be started, but the initial speed is likely to be too high, as shown in figure 8.7.

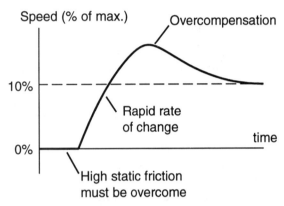

Fig. 8.7 Overcorrection using proportional plus integral control.

An obvious symptom of overcompensation (and of instability in general) is the rapid rate of change of the output speed.

This effect could be detected by subtracting successive feedback signals. If it was found for example that the output speed was increasing too rapidly, the current supplied to the motor could be temporarily reduced, thus tending to oppose the increase in motor speed. Similarly, a rapid decrease in motor speed could be prevented by momentarily increasing the motor current. (The amount of feedback should be less than 100%, so that the rate of change is reduced, rather than reversed.)

In a practical PID control system, the relative amounts of proportional, integral and derivative feedback must be carefully adjusted to provide the ideal response to constantly changing external parameters. Depending on the values chosen, the output response to a step change in the input may be classified as *underdamped*, *overdamped* or *critically damped*, as shown in figure 8.8.

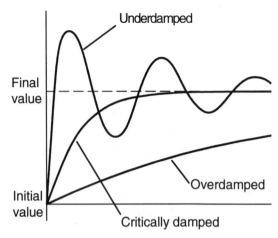

Fig. 8.8 Types of response to a step change in the input.

The ideal response to a step change in the input is where the output moves as rapidly as possible to the final state with little or no overshoot. Underdamped systems tend to oscillate about the final value, while overdamped systems take a long time to settle.

This control systems assignment covers objectives from several level NIII BTEC units, as shown in table 8.1 below.

Unit		Objectives
Microelectronic Systems	E9	Assembly language programs
Electronics	D6	Pulse waveforms
	J12	A – D and D – A conversion
	O18	D.C. motor control
Electrical & Electronic Principles	J15	Electrical machines
	M	Control principles
Mathematics		Calculus
Ind. Control & Instrumentation		Most sections

Table 8.1 Control systems unit coverage.

Integrative Assignment   Page 1 of 1

Control Systems (Ref. NIII/001)

Student: _____   Date:_____

Compet-ence	Description	Claimed grade	Adjusted grade
1	Work within given parameters. Purchase items to budget.		
2	Use a diary, or other method of planning or scheduling tasks.		
4	Transfer skills and knowledge to new situations.		
6	Relate effectively to others.		
7	Allocate individual tasks. Contribute to the group project.		
8	Follow given instructions. Operate equipment according to a manual.		
9	Present information in an appropriate visual form, such as a diagram or plan.		
10	Present information in a written report using appropriate language, tone and layout.		
11	Demonstrate effective verbal communication.		
12	Use sources of further information, such as a library.		
13	Perform experimental procedures.		
14	Work to an action plan.		
15	Demonstrate numeracy (use of a calculator, calculations).		
16	Use technological equipment (computers, instruments, machines).		
17	Demonstrate design skills (creativity, critical review of results).		
18	Use a range of thought processes (logical and creative).		

Signature of learner:_____   Signature of lecturer/supervisor:_____

Fig. 8.9   Common Skills claim form for the control systems assignment.

PLC Design and Construction

Programmable logic controllers (PLCs) are widely used in industry, due to the ease with which they can be programmed and interfaced to industrial equipment.

This assignment considers the design and construction of a simple PLC which is based internally on the *PIC 16C55* microcontroller. The PLC will have up to twelve opto-isolated inputs and eight outputs (which may be relays, Darlington transistors or opto-isolated triacs). Power supply requirements will include +5 and +24 volts d.c. which may either be designed and built as part of the assignment, or may be provided from a separate bench power supply.

The PLC should be constructed using a printed circuit board (using the principles outlined in chapter 6), with the PIC microcontroller held in a *zero insertion force* (ZIF) socket for easy replacement. Programs will be written on a *development system* and stored in the microcontroller's program memory using the supplied EPROM programming unit.

As in chapter five, where the PIC 16C5X series of microcontrollers were introduced, the program development approach proposed here is to use a header file to implement all PIC-related functions. This will then allow programs to be developed using the familiar *ladder logic* approach.

Figure 8.10 shows a block diagram of the PIC 16C55 microcontroller, assuming that ports A and C are used as inputs and port B is configured as outputs.

As an example of the programming method used, consider the *ladder diagram* representation of an SR bistable, as shown in figure 8.11.

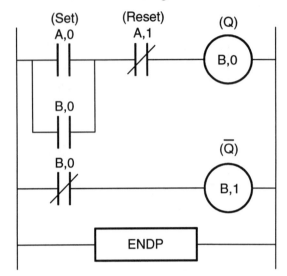

Fig. 8.11 SR bistable represented using a ladder diagram.

The various PLC instruction types are defined as macros in the header file, which allows the ladder diagram to be coded as shown in listing 8.1.

```
        include "plc.h"

        ld          Port_A,0
        or          Port_B,0
        and_not     Port_A,1
        out         Port_B,0
        ld_not      Port_B,0
        out         Port_B,1
        endp
```

Listing 8.1 'PLC' listing for the SR bistable.

The *include* assembler directive causes the appropriate header file to be linked with the assembly language source file during assembly. The content of this header file is given in listing 8.2 opposite.

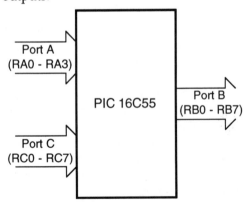

Fig. 8.10 PIC 16C55 external connections.

```
          ;PIC register names
          include "picreg.equ"

          ;PLC command definitions
ld        macro    reg,bit
          movlw    0
          btfsc    reg,bit
          movlw    1
          endm

ld_not    macro    reg,bit
          movlw    1
          btfsc    reg,bit
          movlw    0
          endm

or        macro    reg,bit
          movwf    0fh
          movlw    0
          btfsc    reg,bit
          movlw    1
          iorwf    0fh,w
          endm
or_not    macro    reg,bit
          movwf    0fh
          movlw    1
          btfsc    reg,bit
          movlw    0
          iorwf    0fh,w
          endm

and       macro    reg,bit
          movwf    0fh
          movlw    0
          btfsc    reg,bit
          movlw    1
          andwf    0fh,w
          endm

and_not   macro    reg,bit
          movwf    0fh
          movlw    1
          btfsc    reg,bit
          movlw    0
          andwf    0fh,w
          endm
```

```
out       macro    reg,bit
          movwf    0fh
          btfsc    0fh,0
          bsf      reg,bit
          btfss    0fh,0
          bcf      reg,bit
          endm

endp      macro
          goto     scan
          end
          endm

          ;PLC register names
Aux_1     equ      08h
          ;8 auxiliary relays

          ;Port initialisations
          ;and scan loop
          org      PIC55
          goto     scan

          org      0
scan      movlw    0fh
          tris     Port_A
          ;RA0-RA3 inputs
          movlw    0
          tris     Port_B
          ;RB0-RB7 outputs
          movlw    0ffh
          tris     Port_C
          ;RC0-RC7 inputs
```

Listing 8.2 PLC header file.

This header file provides a relatively simple PLC instruction set which includes the following commands.

- LD   read normally-open contact.
- LD_NOT   read normally-closed contact.
- AND   Logical AND (normally-open).
- AND_NOT   Logical AND (normally-closed).
- OR   Logical OR (normally-open).
- OR_NOT   Logical AND (normally-closed).
- OUT   Send to output.
- ENDP   End of program.

Note that further facilities such as *timers, counters, shift registers* and *drum functions* could also be added to the instruction set by the creation of suitable macro definitions. A further addition might include the use of the *watchdog timer* facility, with the CLRWDT instruction included in the main scan loop to prevent the PLC from 'hanging-up' due to externally-induced electrical noise.

Program statements may refer to the individual bits of ports A, B and C by specifying the name of the port and a bit position (0 = LSB). Eight auxiliary relays have also been provided for temporary variable storage, which is useful when the state of a particular circuit node needs to be stored internally (*block logic* functions for example).

Input and Output Circuit Design

When designing the input and output circuitry, the following points should be noted.

- Screw terminals should be used to allow reliable and safe connection of all input and output signals.
- The logical state of each input or output should be visible during program execution via a series of LEDs.
- Logic 1 and logic 0 inputs should be 24 volts and 0 volts d.c. respectively (which may be provided by a separate d.c. power supply).
- All inputs should use opto-isolators to provide electrical isolation between external voltages and the computer. (Refer to figure 7.24 for a possible input circuit.)
- Each input should be protected from reverse polarity connection using a series power diode.
- Output relays should have a 5 ampere current rating and should be capable of switching d.c. or a.c. loads. (Optionally, several relays may share a single common line.)

- Darlington transistors may optionally be provided in place of relays, on one or more outputs. In this case, the output drive should be through an opto-isolator for safety reasons.
- Opto-isolated triacs may optionally be used in place of relays on one or more outputs. (Refer to figure 7.25 or 7.26 for possible output circuits.)
- Suitable fuses should be placed in series with each set of output contacts (relays, Darlington transistors or opto-isolated triacs). The fuse should have a lower rating than the switching device.
- All circuitry should be enclosed in a suitable insulating box, with only the screw contacts (and the microcontroller ZIF socket) visible to the user.

For safety reasons, it is recommended that the PLC should not be used with mains voltages unless it has been thoroughly tested by a suitably-qualified person. The 24 volt d.c. supply is adequate for many control applications, and low voltage a.c. supplies may be used to test any opto-isolated triac outputs.

Microcontroller Circuit Design

The PIC 16C55 microcontroller requires a small number of external components, in order to function correctly. These include the

- RC or quartz crystal oscillator,
- power on reset circuit.

In addition, the RTCC pin (which is unused in the current design) should be tied to +5 volts using a pull-up resistor (to prevent pick-up of electrical noise).

A possible schematic diagram for the microcontroller, including the above circuits is shown in figure 8.12 opposite.

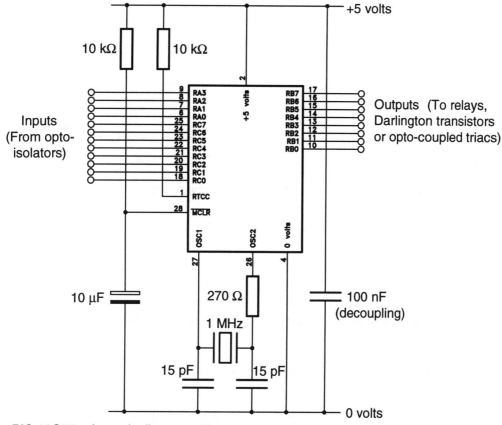

Fig. 8.12 PIC 16C55 schematic diagram with reset and oscillator circuits.

This PLC design and construction assignment covers objectives from several level NIII BTEC units, as shown in table 8.2 below.

A possible common skills claim form for the PLC design and contruction assignment is given in figure 8.13, overleaf.

Unit		Objectives
Microelectronic Systems	E9	Assembly language programs
	F10	Signal degradation (PCB)
Electronics	K	Thyristors
Electrical Applications	H	Switchgear and protection
Microprocessor Interfacing	A	Interface concepts

Table 8.2 PLC design and construction unit coverage.

Integrative Assignment

PLC Design and Construction (Ref. NIII/002)

Student: _____ Date:_____

Compet-ence	Description	Claimed grade	Adjusted grade
1	Work within given parameters. Purchase items to budget.		
2	Use a diary, or other method of planning or scheduling tasks.		
4	Transfer skills and knowledge to new situations.		
6	Relate effectively to others.		
7	Allocate individual tasks. Contribute to the group project.		
8	Follow given instructions. Operate equipment according to a manual.		
9	Present information in an appropriate visual form, such as a diagram or plan.		
10	Present information in a written report using appropriate language, tone and layout.		
11	Demonstrate effective verbal communication.		
12	Use sources of further information, such as a library.		
13	Perform experimental procedures.		
14	Work to an action plan.		
15	Demonstrate numeracy (use of a calculator, calculations).		
16	Use technological equipment (computers, instruments, machines).		
17	Demonstrate design skills (creativity, critical review of results).		
18	Use a range of thought processes (logical and creative).		

Signature of learner:_____ Signature of lecturer/supervisor:_____

Fig. 8.13   Common Skills claim form for the PLC design and construction assignment.

9   GNVQ and Core Skills

Aims

When you have completed this chapter you should be able to:

1 Appreciate the development of the Advanced GNVQ in Engineering/ Technology qualification.
2 Record and develop a profile of Core Skills achievement.

Introduction

At the time of writing, existing *BTEC National level* programmes of study are in the process of being replaced by *Advanced level GNVQ* qualifications. These new schemes are academically equivalent to two 'A' levels, on successful completion, and for this reason are often referred-to as *vocational 'A' levels*.

The *Advanced GNVQ in Engineering/Technology* was piloted from September 1994, and will be nationally available from September 1995. This course, which is nominally of two years duration, consists of 12 units, of which eight are mandatory. The remaining four units are chosen from a wide range of options, thus allowing the student to specialise in a particular engineering discipline. Available areas of specialisation include

- general engineering,
- electrical/electronic engineering,
- mechanical engineering,
- motor vehicle engineering.

Those students with an electrical/electronic bias may select from a range of non-mandatory units including

- communications engineering,
- computer systems and applications,
- electrical principles,
- electrical technology,
- electronics,
- engineering instrumentation and control,
- further mathematics,
- microelectronic systems.

In addition to the twelve engineering units, candidates must demonstrate competence in three mandatory *core skills* units. These are

- communication,
- application of number,
- information technology.

The aim of this section is to provide guidance for those with an interest in the non-mandatory unit *Microelectronic Systems* (which was developed for BTEC by the author). The guidance given here includes

- a cross reference, showing where to find relevent theory in books one and two,
- practical exercises which might form the basis of relevent assignment work,
- activities related to microelectronics, which might provide oportunities for the claiming of mandatory core skills.

Microelectronic Systems Unit

It should be remembered that, due to the pilot nature of the Advanced GNVQ in Engineering/ Technology, detailed unit contents are likely to change from time to time. Bearing this in mind, the following cross-referenced tables should still be useful when particular information needs to be located (although the latest version of the unit specification should always be used).

The unit consists of the following elements, each of which has several performance criteria and a statement of the required range.

- Investigate features of microprocessor based sytems.
- Prepare software to solve an engineering problem.
- Investigate the interfacing of a microprocessor based system.

Tables 9.1, 9.2 and 9.3 suggest where relevent information may be located in *Microelectronic Systems – One* and *Two*, for each of the above elements.

Range	Where to look...
Microprocessor and computer	Book one – 2 Microprocessor based systems. Book two – 1 Microprocessor based systems. Book one – 3 The fetch execute cycle. Book one – 4 Microprocessor instruction sets. Book one – 7 Subroutines and the stack. Book two – 3 The stack and interrupts.
Design factors and trends	Book two – 5 Trends in microprocessor design. Book two – 2 Memory devices.

Table 9.1 Features of microprocessor based systems.

Range	Where to look...
Facilities	Book one – 5 Designing and testing programs. Book one – 8 Practical applications. Book two – 7 Practical applications.
Structured design techniques	Book one – 5 Designing and testing programs. Book one – 6 Assembly language programming.
Instruction types	Book one – 4 Microprocessor instruction sets.

Table 9.2 Prepare software to solve an engineering application.

Range	Where to look...
Types of interface	Book one – 5 Designing and testing programs. Book one – 8 Practical applications. Book two – 7 Practical applications.
Interface protocol	Book one – 5 Designing and testing programs. Book one – 8 Practical applications. Book two – 7 Practical applications.
Documentation	Book one – 4 Designing and testing programs. Book one – 6 Assembly language programming. Book two – 6 Principles of system design.

Table 9.3   Investigate the interfacing of a microprocessor based system.

Candidates must provide evidence showing that the performance criteria associated with each element have been achieved. Tutors must therefore devise suitable assessment opportunities which will allow each of the elements to be demonstrated by the candidate.

On such programmes, there is clearly a need to minimise the assessment workload of each candidate. This may be achieved through careful assignment design, in order to maximise the number of elements and core skills which may be claimed from each assessment. There is no reason why a single assignment should not cover all of the microelectronic systems performance criteria and also several common skills. Two or more of such assignments would then give candidates multiple opportunities for claiming each element.

The 70 practical exercises in books one and two form a useful resource, which may be used as the basis for assignment work design. Due to the reduced learning support time, and the more general nature of GNVQ qualifications, it is suggested that chosen topics should be kept relatively simple. Bearing this in mind, table 9.4 suggests practical exercises which might be relevent to the three microelectronic systems elements.

Finally, tables 9.5, 9.6 and 9.7 (overleaf) suggest relevent activities which might form a basis for claiming the three mandatory core skills from microelectronics assignment work.

Element	Relevant exercises
Microprocessor based systems	Book one – Microelectronics revolution. (Most practical exercises where an analysis of the required hardware and software is performed.)
Software development	Book one – Mathematical subroutines, time delays, stopwatch program.
Interfacing	Book one – transistor as a switch. Book two – waveform generation, electronic dice.

Table 9.4   Relevant practical exercises.

Core skill	Activity
Take part in discussions	Discuss methods of tackling problems. Explain operation of working system. Allocate tasks in group work.
Prepare written material	Reports, program listings, test results.
Prepare images	Circuit diagrams, flowcharts, structure diagrams.
Respond to written material and images	Follow instructions in assignments. Refer to notes and manufacturers data sheets.

Table 9.5   Communication core skills.

Core skill	Activity
Gather and process data	Measurement of program execution time. Recording of actual program output.
Represent and tackle problems	Calculation of program execution time. Calculation of predicted program output.
Interpret and present data	Comparison of actual and predicted data (time delay, output from mathematical software, statistical analysis of electronic dice output).

Table 9.6   Application of number core skills.

Core skill	Activity
System set-up, input data	Text editor (entry of assembler source code), word processor (reports), CAD (diagrams).
Edit, organise and integrate information from various sources	Assembler 'include' files, use of illustrations in word processed or desktop published reports.
Select and use presentation formats	Program listings, layout of reports, labelling of illustrations.
Evaluate features of existing software	Assembler, simulator, debugger, monitor, downloader, operating system.
Deal with errors	Syntax errors, logical errors, hardware errors, equipment errors.

Table 9.7   Information technology core skills.

10 Answers to Problems

Chapter 1

1 a) A *microprocessor* is a single IC containing the major elements of a microcomputer, including registers, the control unit and the arithmetic and logic unit (ALU).
   b) A *microcomputer* is a computer whose central processing unit (CPU) is a microprocessor.
   c) The *CPU* of a microcomputer is a microprocessor, although more complex computers may use several components for the same function.
   d) *Registers* are storage locations which are internal to the microprocessor. Many registers have special functions.

2 The *address bus* selects a single memory or I/O port location for reading or writing of data. (8 bit microprocessors normally have a 16 line address bus). Once the required location has been selected using the address bus, the data is transferred to or from the microprocessor via the *data bus*. (8 bit microprocessors have an 8 bit data bus.) The *control bus* is a miscellaneous group of wires which allows the microprocessor to control and synchronise the actions of each part of the microcomputer.

3 a) The *accumulator* is a general purpose register which is normally loaded with the result of any arithmetic or logical calculations performed by the ALU.

   b) The *program counter* register contains the address of the next byte which is to be loaded during execution of the machine code program.
   c) The *flags register* or *processor status register* contains a series of individual flags (individual bits) which indicate the current state of the microprocessor. The operation of certain instructions may be altered, based on the state of particular flags.
   d) The *stack pointer* is closely related to an area of RAM called the stack, which is used for temporary data storage. During data storage or retrieval, the stack pointer is either incremented or decremented so that it always points to the top of the stack. The stack is therefore called a *last-in, first-out* store.

4 *Op. code fetch.* The retrieval from memory of the operation code associated with the next instruction in the program. Active control bus signals are $\overline{M_1}$, $\overline{MREQ}$ and $\overline{RD}$.
   Memory read. The retrieval from memory of a single byte of data. Active control bus signals are $\overline{MREQ}$ and $\overline{RD}$.
   Memory write. The storage in memory of a single byte of data (from a microprocessor register). Active control bus signals are $\overline{MREQ}$ and $\overline{WR}$.
   Input/output read. The retrieval from an input port address of a single byte of data. Active control bus signals are $\overline{IORQ}$ and $\overline{RD}$.

Input/output write. The storage at an output port address of a single byte of data (from a microprocessor register). Active control bus signals are $\overline{IORQ}$ and $\overline{WR}$.

5 *Op. code fetch*. The retrieval from memory of the operation code associated with the next instruction in the program. Active control bus signals are SYNC, R/$\overline{W}$ (at logic 1), ϕ_1 and ϕ_2.

Memory read. The retrieval from memory of a single byte of data. Active control bus signals are R/$\overline{W}$ (at logic 1), ϕ_1 and ϕ_2.

Memory write. The storage in memory of a single byte of data (from a microprocessor register). Active control bus signals are R/$\overline{W}$ (at logic 0), ϕ_1 and ϕ_2.

6 If several devices, capable of outputting data, are connected to the data bus, only one device will attempt to control the bus at any time. The microprocessor, via its control bus and associated address decoding circuitry, ensures that only one device is allowed to enable its outputs. All other devices connected to the bus place their outputs in a high impedance state. This prevents electrical contention where two or more devices try to impose differing logic levels on the same bus line.

7 Yes. The maximum current demanded by the five inputs is 2.5 mA, while the output can only supply 2 mA. The excessive current drain would either cause improper logic levels or lead to overheating and possible electrical failure of the output circuit.

8 *Address decoding* circuits convert the state of the most significant address bus lines into individual chip enable signals, thus ensuring that only one memory or input/output device is enabled at any time.

Using a single address line (A_{15}) either of two 32 kilobyte memory blocks may be selected. The use of two address lines allows one of four 16 kilobyte memory devices to be enabled, while three address lines decode up to eight memory blocks, each of 8 kilobytes.

The system *memory map* is determined by the chip enable signals used and the type of memory (RAM or ROM) fitted in each memory block.

Chapter 2

1 a) NPN. $V_B = V_E + 0.7$ volts.
   b) PNP. $V_B = V_E - 0.7$ volts.
   c) N-channel enhancement MOSFET.
      V_{GS} positive.
   d) N-channel depletion MOSFET.
      V_{GS} zero or positive.
   e) P-channel enhancement MOSFET.
      V_{GS} negative.
   f) P-channel depletion MOSFET.
      V_{GS} zero or negative.

2 Refer to the appropriate sections of chapter 2 for descriptions of the internal construction and operation of TTL, NMOS, CMOS and ECL circuits.

3 *Mask programmed ROM* is the most cost effective form of non volatile memory in volume production situations. The ROM content is fixed at the final stage of manufacture when a customer specific 'mask' is overlaid onto the silicon slice. This determines which of the transistors in the memory array are connected or left unconnected, hence determining the logic level of each individual bit.

EPROM or *erasable programmable ROM* may be (electrically) programmed using an EPROM programmer, and may be erased by exposure to ultraviolet light, via a transparent window. The relatively low cost of EPROM programming equipment makes EPROMs a cost effective form of non-volatile memory, in small scale production situations.

EEPROM or electrically erasable programmable ROM is programmed in a similar way to an EPROM, but without the need for externally applied programming voltages. The programming process may also be reversed, eliminating the need for erasure using ultraviolet light. These features allow EEPROMs to be programmed many times without removal from the socket.

4   *Static RAM* operation is based on the bistable circuit, in which a pair of cross-coupled transistors are used to store a single bit of information.. Since one half of the circuit always conducts, power consumption is often greater than dynamic RAM. Low power CMOS static RAMs are also available which are suitable for battery-backed (non volatile) operation.

Dynamic RAM uses the presence or absence of stored charge to represent a single bit of data. Because of charge leakage it is necessary to refresh dynamic RAMs regularly to retain their data content. Due to the simple construction of dynamic RAM, they are available in higher memory capacities than static RAM, although this advantage is offset by the need for additional support circuitry.

5   a) *CMOS static RAM* with battery backup would be suitable for storing valuable data, which might otherwise be lost due to a power failure. The advantage of non-volatile RAM over EEPROM in this application is the faster write time (100 ns compared with 2 ms) and the infinite number of possible write cycles.

b) *Mask programmed ROM* is ideal for use in large scale mass production (where the ROM content is certain to remain fixed), due to its low unit cost.

c) EPROM is suitable for small scale production, particularly where the ROM content may need to be updated at some time in the future. EPROM programming equipment is relatively cheap and easily available, making EPROMs ideal in prototype development situations.

d) *OTP (one time programmable) ROM* is normally based internally on EPROM technology, but the transparent erasure window is omitted. This reduces production costs, while allowing devices to be programmed in the normal way. The reduced cost makes this device suitable for small scale production runs, where erasure will not be required.

e) *EEPROM* is the one of the most flexible forms of non volatile storage, due to its ability to be programmed without additional equipment. This makes EEPROMs suitable for applications where the ROM content may need to be periodically updated under program control. A possible application would be the storage of telephone numbers in a telephone memory.

6   a) After 10 years, 50000 electrons have escaped. Approximately 13.7 electrons escape each day, on average. (Roughly one electron per hour!)

b) The average current which flows over the ten year period is 2.5×10^{-23} amperes (or coulombs per second).

7   a) 5 seconds.
    b) 205 seconds.

Chapter 3

1 a) A *subroutine* is a mini-program which performs some distinct function and which can be called from the main program as often as required. An area of memory called the *stack* is used to hold the subroutine's *return address*, which is the address of the next instruction to be executed in the main program.

   b) A *nested subroutine* is a subroutine that is called from within another subroutine. Control is returned to the calling subroutine before being passed to the main program. Nested subroutine operation is possible due to the *last-in, first-out* nature of the stack.

   c) The *stack* is an area of RAM that is reserved for the storage of temporary data, such as subroutine return addresses or the content of data registers.

   d) The *stack pointer* is a microprocessor register that points to the top of the stack. This may be either the last used location or the first unused location, depending on the type of microprocessor. As data is stored on the stack, the stack pointer decrements, thus causing the stack to grow in size. Data removal causes the stack pointer to increment, reducing the size of the stack. The stack is said to be a *last-in, first-out* store, since the most recently saved data is always on the top of the stack.

2 When a jump to a subroutine is made, the address of the instruction immediately following the subroutine call is pushed onto the stack. This 16 bit value is stored as two bytes with the most significant byte stored first. The stack pointer register is decremented twice during the storage of the return address. The address of the subroutine is then placed into the program counter register, causing the subroutine to be executed.

The last instruction in the subroutine is a *return from subroutine*, which causes the top two bytes to be popped from the stack and placed into the program counter. The stack pointer register is incremented twice during this process. The main program then resumes at the next instruction after the jump to subroutine.

3 A *push* instruction may be used to place the content of a specified register on the stack. This register may then be altered as required and later restored to its previous value using a *pull* or *pop* instruction.

   Using a series of push instructions, it is possible to save all of the data registers on the stack. As the stack operates as a last-in first-out store, the data must be later removed from the stack in reverse order.

   By using the stack to save any registers which are used by a subroutine, prior to performing the subroutine function, it is possible to return all registers to their previous state on completion of the subroutine.

4 a) An *interrupt* is an externally generated signal, which is applied to the microprocessor. If the interrupt is accepted, execution of the main program is temporarily suspended, allowing the external device to be serviced, prior to resuming the main program.

   b) A *nested interrupt* is an interrupt which occurs during the processing of another interrupt. The previously executing interrupt service routine is temporarily suspended, allowing the new interrupt to be dealt-with. On completion, the previous ISR continues from the point reached, after which (in the absence of further interrupts), the main program resumes.

   c) *Maskable interrupts* are externally generated requests for attention, which may be either accepted or ignored by the microprocessor under program control. The programmer must *enable interrupts* before using maskable interrupts.

d) A *non maskable interrupt* is a high priority interrupt which cannot be ignored by the microprocessor. This type of interrupt is generally reserved for important external events which the microprocessor must not ignore. Examples might include early warning of power supply failure or an alarm condition.

e) An *interrupt service routine* is a mini-program which is executed in response to an externally generated interrupt. ISRs are very similar to subroutines and they should leave all registers used by the main program unaltered, so that the main program can later resume without error. As with subroutines, push and pop (or pull) instructions may be used for this purpose.

f) In a system which uses *vectored interrupts,* the interrupting device supplies the address of the ISR which must deal with the interrupt. This contrasts with systems where the ISR is always stored at a particular location in memory. Vectored interrupts are particularly suitable in situations where multiple interrupts are used, as each device may have its own ISR.

5 A non maskable interrupt is triggered by a logic 1 to logic 0 transition on the NMI pin. In response to the interrupt, the currently executing instruction in the main program is completed, after which the program counter register is pushed onto the stack. (The 6502 also stores the flags register on the stack.) A jump is then made to the non maskable interrupt service routine.

Maskable interrupts are automatically disabled during the ISR. (The mechanism used varies from one microprocessor to another.)

The ISR itself should preserve all registers used by the main program and this is normally achieved by pushing all registers onto the stack at the start of the ISR. The same registers must be popped from the stack at the end of the ISR, in reverse order.

A return from interrupt (the Z80 uses different return instructions for maskable and non maskable interrupts) instruction causes the ISR return address to be popped from the stack. The main program then resumes from the point reached.

6 In a *polled* system, the state of external devices is regularly tested by the microprocessor. This takes place during the main program which normally executes in a continuous loop. The microprocessor spends a lot of time testing devices which do not require attention, which is rather wasteful. The response time to an external event may also be slow in some cases, since it depends on the point reached in the polling loop and its speed.

An *interrupt* driven system allows the main program to execute normally until an interrupt is detected. At this point the main program is temporarily suspended and an ISR is called to deal with the external event. Interrupts allow the microprocessor to provide a faster response to external events then polling. No time is spent testing devices which do not require attention.

7 In a system which is connected to multiple interrupt sources, some interrupts are likely to be more important than others. If two or more interrupts occur simultaneously, the microprocessor must deal with important interrupts first, before considering lower priority requests for attention.

In addition, an interrupt priority system should ideally allow a partially completed ISR to be temporarily suspended, if a higher priority interrupt occurs before its completion.

8 Four methods of prioritising interrupts are considered in the main text. These are

- testing of status flags in software,
- priority encoder,
- daisy chain system,
- interrupt priority register.

Please refer to the main text for a description of each technique.

Chapter 4

1 Handshaking allows the sender and receiver to control and synchronise the flow of information across an interface. In hardware based handshaking systems, a number of extra wires are used for control purposes, while in software based arrangements, special characters may be used to indicate device status.

When handshaking is used, data communication may proceed as quickly, or as slowly, as the situation allows. With a computer to printer connection, initial data transfer may be at high speed, with incoming data temporarily held in the printer buffer. If the buffer is filled, then further data may only be transmitted as space becomes available in the buffer. Once all data has been placed in the buffer, the computer is then free to continue with other tasks. The interrupt driven nature of hardware based handshaking systems also means that both computer and printer can perform other tasks when the interface is idle.

2 The Centronics interface allows 8 bit parallel data communication, most commonly between a computer and printer. Handshaking lines are used to indicate device status. In particular a 1 to 0 transition on the STROBE line signals that valid data is present on the data outputs, while a similar transition on the ACK (acknowledge) pin indicates data reception. The printer may also indicate its general status by asserting the BUSY signal (when the printer buffer is full, or the paper tray is empty for example).

3 Parallel data communication has the obvious advantage that it is inherently faster than serial data transfer, due to the simultaneous transmission of several data bits. This same feature also results in higher cabling costs for parallel interfaces, due to the increased number of conductors. The use of TTL voltage levels by parallel interfaces (which has relatively low noise immunity), together with the capacitive coupling between adjacent conductors, effectively limits the length of parallel interface cables to about 2 metres. The RS232 serial interface uses special voltages for data signals (with higher noise immunity), allowing reliable data communication over much longer distances. Effectively, the maximum cable length is limited by the maximum allowed capacitance of the line, divided by the cable capacitance per metre (typically around   15 metres). Even greater cable lengths are possible if the baud rate is reduced.

4 Refer to the practical exercise on page 124 for a typical shift register circuit diagram.

Incoming serial data may be converted to parallel form by applying clock pulses to the shift register at regular intervals of time. Each clock pulse causes the data present on a particular output to be shifted one place to the right. After the appropriate number of pulses, the data may be read from the bistable outputs in parallel form. This is commonly referred to as *serial in, parallel out (SIPO)* operation.

The creation of serial output data is the inverse of the above process. The register is initially loaded with parallel data (using suitable gating logic) so that each bistable holds a single data bit. A series of clock pulses then cause the data to be shifted out in serial form. This is commonly referred to as *parallel in, serial out (PISO)* operation. (In a practical serial interface, additional information such as start, stop and parity bits may also be added and later tested by the interface hardware.)

5 The RS232 interface is a widely used serial interface, which was originally developed to define the electrical connections between a computer (the *data terminal equipment* or *DTE*) and *modem* (the *data communication equipment* or *DCE*). The RS232 standard defines a voltage in the range −5 to −25 volts at the transmitter to be a logic 1, while voltages between +5 to +25 volts indicate a logic 0. At the receiver, any voltage greater than +3 volts is accepted as a logic 0, while voltages less than −3 volts are recognised as logic 1. The use of special voltage levels requires the use of level shifting hardware at either end of the interface to convert signals to and from TTL levels. It also results in greater noise immunity and more reliable data transfer. In addition to the transmitted and received data lines (TXD and RXD), a number of other signals are also used to establish the connection. Refer to table 4.9 in the main text for a description of these signals.

6 A binary number is said to have even parity if it has an even number of ones (0, 2, 4 etc.) and odd parity if it has an odd number (1, 3, 5 etc.) of ones. By adding an extra parity bit during data transfer, it is possible to ensure that each transmitted data word has the chosen type of parity. Should a single bit data error occur during transmission, this will inevitably lead to a change in parity, which may be detected at the receiver. The receiver may then request a retransmission of the erroneous data. Parity bits provide a relatively reliable method of detecting data errors, provided the probability of an error is small. This is because some multiple bit errors may be undetected by the parity circuitry. See the practical exercise on page 122 for parity generation and checking circuits.

7 a) Start bit. In asynchronous serial data transfer, this bit is used as a timing reference for the following data bits. Each bit is sampled at its midpoint by the receiver to reduce the chance of errors caused by differences in transmitter and receiver clock frequencies.

b) Stop bit. One or more bits which follow the data bits. If an invalid logic level is detected at this point, this may indicate a data framing error.

c) Even parity. A data word has even parity if an even number of bits are at logic 1. Parity is normally used by error detection circuitry associated with the interface.

d) Null modem cable. This is a special purpose RS232 interface cable which allows two DTEs (or DCEs) to be linked via a serial interface. Data pins 2 and 3 are normally swapped and other connections may also be necessary to allow reliable communication.

8 Real time clock (RTC) ICs are used in computers where the current date and time must be made available. This includes computers whose operating system software stores date and time information in association with program and data files when they are created or altered. Control systems which must perform tasks at particular times may also use a real time clock. The RTC appears to the programmer as a series of registers (year, month, day, hours, minutes, seconds and fractions of a second) which may be read or altered under software control. These registers are regularly updated, even when the computer itself is switched off, due to the use of a battery backed power supply.

9 Counter/timer ICs may be used to:

a) Measure a single time interval.
b) Count externally applied pulses.
c) Generate interrupt requests at regular intervals.

Chapter 5

1 Data and address bus widths have steadily increased since the introduction of the microprocessor in 1971. Increased address bus widths allow the microprocessor to physically address larger quantities of memory, which in turn is a response to the increasing complexity of modern software.

The width of the data bus has tended to increase along with the internal register widths of modern microprocessors. This allows registers to be loaded in a single memory read operation. Another reason for this increase is that a single memory read operation may be used to read several successive memory locations, allowing the microprocessor's instruction queue to be filled in a single operation.

Increases in clock frequency have allowed microprocessors to perform more instructions in a given period of time, thus increasing the overall performance of microprocessors.

Each of these trends is likely to continue for the foreseeable future.

2 a) The use of *Virtual memory* allows programs that are larger than the physical computer memory to be executed by the microprocessor. During program operation, only the active section of the program is loaded from backing store. When an attempt is made to access an area of the program that is not currently available, a type of error called an *exception* is generated. This causes the required program section to be loaded, after which program execution continues.

b) *Multitasking* is a time and resource sharing technique, which allows a microprocessor to give the appearance of performing several simultaneous tasks. In fact, each program runs for a short interval of time, on a repeating basis.

c) Multiplexed bus arrangements are used by some microprocessors to minimise the number of external connections required by the microprocessor. With the Intel 8085 and 8086 microprocessors, certain pins act as address or data pins at different times in the instruction cycle. External latch ICs are normally required to demultiplex these signals.

d) *RISC* or *reduced instruction set* microprocessors typically have a small instruction set, which is enhanced to provide high speed program execution. This contrasts with traditional or *complex instruction set* microprocessors (*CISC*), which tend to execute programs more slowly.

e) *Superscalar* microprocessors have multiple execution units, allowing several operations to be performed simultaneously. Special logic inside the microprocessor is able to identify instructions in the instruction queue which do not conflict, and these are then passed to their own execution unit. (An increased data bus width allows the instruction queue to be rapidly filled from memory, as mentioned earlier).

f) *Harvard architecture* microprocessors use separate buses to access program and data memory (and may also use separate memory banks for these two functions). This allows the efficiency of the microprocessor to be increased by instruction pipelining, where the execution phase of one instruction overlaps with the fetch cycle of the next.

3 The Intel 8086 microprocessor consists of several sub-sections, each of which contribute to the overall efficiency of operation.

a) The *execution unit* executes instructions which are taken from the *instruction queue*.

b) The *bus interface unit* handles all communication between the microprocessor and external devices, and ensures that the instruction queue is kept as full as possible at all times.

c) The *instruction queue* is filled by the bus interface unit, which operates independently of the execution unit. The execution unit repeatedly removes instructions from the instruction queue, with subsequent instructions moving-up toward the top of the queue. In the case of a jump instruction, the queue is flushed and the bus interface unit begins filling the queue from the new program address.

d) The Intel 8086 microprocessor uses a *segmented memory system* in which a physical address is formed by combining (adding) the content of an *offset* and a *segment register*. Each segment is up to 64 kilobytes in size and can reside anywhere within the 1 megabyte address space. The segment register provides the most significant 16 bits of the address, while the offset gives the lower 16 bits.

4 A *microcontroller* contains the main features of a microcomputer, housed within a single integrated circuit. Typically this includes the microprocessor, RAM, ROM, input/output ports, real time clock/counter, watchdog timer, and in some cases analogue signal input/output.

These devices combine a high degree of functionality with small size and low cost, making them ideal for *embedded control* applications. Their ability to be reprogrammed greatly simplifies the task of modifying existing designs, thus reducing costs while increasing design flexibility.

Chapter 6

1 Several methods of producing single sided PCBs exist. The most commonly used technique is based on simple photographic principles.
In this case the PCB is coated with a light sensitive emulsion, which is developed by exposure to ultraviolet light inside a sealed light exposure unit.

A transparent acetate sheet, containing a positive image of the desired track layout is placed over the photo-sensitive PCB during exposure. Unwanted areas of the board are then developed by exposure to the ultraviolet light.

The photo-sensitive coating in these developed areas is then stripped away by immersing the PCB in a dilute solution of sodium hydroxide. At this point the copper-clad side of the PCB is covered by a positive photographic image of the desired track pattern. Unwanted areas of copper are then stripped away by immersion in a ferric chloride solution, after which the board is thoroughly washed.

Suitably-sized holes are then drilled in the PCB, after which components may be inserted and soldered in place.

2 Available PCB types include

- Single sided,
- Double sided,
- Double sided with through-plated holes,
- Multiple layer,
- Flexible.

Single sided PCBs are suitable for simple circuits and may easily be manufactured by the amateur.

Double sided boards are used with more complex designs, in preference to the use of wire links on the component side of the PCB. Great care must be exercised to ensure correct alignment between pads on each side of the board. This is normally achieved by using a double sided light exposure unit.

Connections between different sides of the PCB may be made using soldered pins or by soldering component leads on both sides of the board. With professional PCB production, through-plated holes are normally used to link tracks on either side of the PCB, thus simplifying the PCB assembly process.

Multiple layer boards are used in professional designs where the component packing density is very high. Boards containing surface mount components are a typical example.

Flexible PCBs are used in compact designs and allow the normally bulky PCB connectors to be eliminated. Their inherent suppleness is also useful where the product itself must posses some degree of flexibility. Portable consumer electronic products, are ideal applications for flexible PCBs.

3 A possible PCB layout, based on the connection matrix, is shown below.

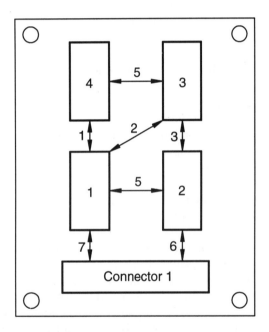

4 The resistance of the track is given by

$$R = \frac{\rho \, l}{A}$$

$$= \frac{0.0173 \times 10^{-6} \times 0.2}{0.035 \times 10^{-3} \times 0.25 \times 10^{-3}}$$

$$= 0.4 \ \Omega$$

5 a)

$$C = \frac{\varepsilon_0 \, \varepsilon_r \, A}{d}$$

$$= \frac{8.85 \times 10^{-12} \times 5.0 \times 0.2 \times 0.05}{1.5 \times 10^{-3}}$$

$$= 0.3 \ nF$$

b) Capacitive coupling between power supply tracks is actually an advantage. This has the same effect as decoupling capacitors, which are used to combat power supply and track inductance.

6 Careful PCB layout may be used to minimise unwanted circuit effects such as track resistance and inductive or capacitive coupling between tracks. All three effects are minimised by minimising the length of PCB tracks. In addition,

- increasing track width reduces the track resistance for a given track length (while tending to increase its capacitance). Power supply and ground track widths should also be increased (compared with signal track widths), due to their larger current capacity.
- reducing the area of current carrying loops minimises the production of magnetic fields due to changing current. Loops which do not generate magnetic interference, tend not to receive it, because the mutual inductance between loops is minimised.

  Placing power supply and ground tracks on opposite sides of the PCB is a useful technique for minimising loop area, as is the earth plane.
- Capacitive coupling can be minimised by keeping conductors as far apart as possible, while minimising the area of the tracks. Deliberately providing capacitive coupling to ground is also effective.

7 a)

$$E = L \frac{di}{dt}$$

$$= 10 \times 10^{-9} \times 4 \times 10^{9}$$

$$= 40 \text{ V/cm}$$

b) An obvious defence is the use of *clamping diodes*. Microprocessors may also be fitted with a *watchdog timer* circuit in electrically noisy environments.

8 The current consumption of digital ICs rises momentarily during switching. This change in current can lead to induced e.m.f.s on the power supply rails, due to track and power supply inductance.

Decoupling capacitors may be used to supply these momentary current surges. A single large capacitor is normally placed close to the point where d.c. power is supplied to the PCB. Smaller decoupling capacitors are then placed in close proximity to individual ICs, with typically one IC for every two ICs. Typical values for the PCB decoupler are 10–100 μF, with 0.1 μF ceramic capacitors being used for IC decoupling.

9 a)

$$Z_c = \sqrt{\frac{L}{C}}$$

$$= \sqrt{\frac{6 \times 10^{-9}}{0.5 \times 10^{-12}}}$$

$$= 110 \ \Omega$$

b) The effects of signal reflections or ringing may be reduced by minimising the length of PCB connections, and hence the time taken for the signal to travel along the track. Signal reflection becomes a problem when the propagation time is greater than on third of the signal rise or fall time.

Unwanted signal reflections can be avoided by terminating a track with a load impedance which is equal to the *characteristic impedance* of the transmission line formed by the PCB connection.

The load impedance may be modified using *bus termination* circuits, although these increase the component count and in some cases degrade the electrical characteristics of the signal. Alternatively *line driver* ICs may be used, whose output and input impedances are matched to that of the transmission line.

Another technique is to use *clamping diode* arrays at the receiving end, which are effective against voltage transients and crosstalk.

10 a) Five sources of (externally generated) electrical noise are

- electrostatic discharge (ESD),
- electromagnetic pulse (EMP),
- radio frequency interference (RFI),
- earth noise,
- earth loop.

b) *Clamping diodes* are a useful defence against *electrostatic discharge* (which is caused by static build-up on an object or person, followed by sudden discharge).

In general, if the noise arrives at the victim circuit directly through the electrical wiring (known as *conducted* noise), the cure is to use *filters* or *suppressors*. *Decoupling capacitors* and clamping diodes are examples of this type of protection.

If the noise arrives in the form of electromagnetic radiation (*radiated* noise), then *shielding* techniques are normally used. These include the use of *Faraday shields* as a defence against electric fields and the minimisation of *loop area* in the case of inductive coupling.

Careful earth layout design is the main defence against *earth noise* and *earth loops*.

Index